Study Guide to Accompany

UNDERSTANDING HUMAN ANATOMY AND PHYSIOLOGY

Study Guide to Accompany

UNDERSTANDING HUMAN ANATOMY AND PHYSIOLOGY

Eldra Pearl Solomon

P. William Davis
Hillsborough Community College

McGRAW-HILL BOOK COMPANY
New York   St. Louis   San Francisco   Auckland   Bogotá   Düsseldorf
Johannesburg   London   Madrid   Mexico   Montreal   New Delhi
Panama   Paris   São Paulo   Singapore   Sydney   Tokyo   Toronto

Study Guide to Accompany
UNDERSTANDING HUMAN ANATOMY AND PHYSIOLOGY

Copyright © 1978 by McGraw-Hill, Inc.
All rights reserved.
Printed in the United States of America.
No part of this publication may be reproduced,
stored in a retrieval system, or transmitted,
in any form or by any means,
electronic, mechanical, photocopying, recording, or otherwise,
without the prior written permission of the publisher.

ISBN 0-07-059647-6

1234567890 DODO 78321098

The editors were James E. Vastyan, Janet Wagner, and Anne T. Vinnicombe;
the production supervisor was Charles Hess.
The drawings were done by Lou Barlow, Charles H. Boyter,
D. L. Cramer, Carol Donner, Gottfried W. Goldenberg,
Marcia Williams, and J & R Services, Inc.
R. R. Donnelley & Sons Company was printer and binder.

Cover: Muscles according to the eighteenth century anatomist Albinus.
*(From Albert Bettex, "The Discovery of Nature." By permission of
Droemersche Verlagsanstalt and Simon & Schuster, Inc., New York.)*

CONTENTS

1. INTRODUCTION TO THE BODY — 1
2. CHEMISTRY AND LIFE — 14
3. THE CELL: *Basic Unit of Life* — 30
4. THE INTEGUMENTARY SYSTEM — 49
5. THE SKELETON — 65
6. THE MUSCLES — 81
7. THE MUSCULOSKELETAL SYSTEM — 91
8. THE NERVOUS SYSTEM: *Basic Organization and Neurophysiology* — 123
9. THE NERVOUS SYSTEM: *Divisions and Functions* — 142
10. THE SENSE ORGANS — 166
11. PROCESSING FOOD: *The Gastrointestinal System* — 186
12. UTILIZATION OF NUTRIENTS: *Metabolism* — 209
13. THE CARDIOVASCULAR SYSTEM — 231
14. THE LYMPHATIC SYSTEM — 256
15. THE RESPIRATORY SYSTEM — 274
16. THE URINARY SYSTEM — 295
17. ENDOCRINE CONTROL — 318
18. REPRODUCTION — 340
19. DEVELOPMENT AND GENETICS — 371

USING THE STUDY GUIDE

This study guide has been designed to help you identify concepts to be learned, to learn them, and to test your mastery of them. Students who diligently use such a study guide generally benefit from it considerably, but the key word here is "diligently".

1. First read the chapter in the textbook, using the learning objectives as a guide to tell you what you should be able to do when you have mastered the material. We suggest that you underline, or highlight in color, all apparently important material with emphasis upon basic facts and concepts. It is also particularly helpful to keep a list of all new terms and definitions.

2. Take the *pretest* at the beginning of the corresponding chapter in the Study Guide. Answers are given at the end of the pretest, with references to specific learning objectives. Your answers need not be precisely those given, but they should be close in meaning. The purpose of the pretest is to give you some indication of how well you have mastered each learning objective and in which of them you are weakest.

3. You may choose at this point to simply go through the entire study guide chapter, and this is probably the best thing to do for the sake of review. If, however, you are short of time, find the LO section in the Study Guide chapter which the pretest indicates you should study further. (Study Guide chapters are arranged into sections by LO number). Study the summary statement under the LO, and if necessary, study the appropriate section of the textbook. Perform the tasks (fill in the blanks, label diagrams, etc.) called for in the LO section.

4. Answer the *test-yourself* questions at the end of the LO section in which you have been working. Check your answers against the correct answers given at the end of the study guide chapter. If you are wrong, try to discover why by careful rereading of the appropriate material in the textbook.

5. When you think you have mastered all the LOs, take the *post test* at the end of the study guide chapter. An answer key to these is also provided.

Even after each chapter exercise has been completed, it may be useful to retake the various tests as part of a review for a cumulative test or final examination. Many instructors deliberately include some of the questions from study guides such as this in their classroom testing program.

Eldra Pearl Solomon
P. William Davis

1 INTRODUCTION TO THE BODY

PRETEST

1. The study of the effect of disease processes upon the human body structure and function is called __pathology__

2. Physiology is the study of body __function__

3. The study of normal tissues is known as __histology__

4. The foot is (**inferior**/superior) to the knee but it is also __distal__ to the knee

5. The shoulder is __lateral__ to the breastbone

6. The breastbone is __anterior__ to the spine

7. The stomach is located in the __abdominal__ cavity

8. The main organs located in the thoracic cavity are the __heart__ and the __lungs__.

9. The tendency of the body to resist harmful change is called __homeostasis__

10. In __negative feedback__, a product inhibits the rate of its own formation.

11. Regulations of the chemistry of the blood is mostly the responsibility of the __urinary__ system

12. The _____ system consists of ductless glands

13. The most powerful muscles moving the thumb are its _____ muscles

14. These muscles exert pull upon the bones of the thumb by means of cable-like _____

15. Epithelial tissue _____ the body surface and _____ many hollow organs

16. The liver is a member of the _____ system

17. All epithelial tissue possesses a thin, non-cellular _____

ANSWERS: 1.pathology (LO 1) 2.function (LO 2)
3.histology (LO 1) 4.distal*(LO2) 5.lateral (LO 2)
6.anterior (LO 2) 7.abdominal (LO 3) 8.heart, lungs
(LO3) 9.homeostasis (LO 4) 10.negative feedback (LO 4)
11.urinary (LO 5) 12.endocrine (LO 5) 13.extrinsic
(LO 6) 14.tendons (LO 6) 15.covers, line (LO 4)
16.gastrointestinal (digestive)(LO 7) 17.basement membrane (LO 7)

*and, at the same time, inferior

LO 1: List and be able to define the major
branches of anatomy and physiology to be
discussed in this book.

Anatomy deals with body structure, physiology with function and pathology with disease processes. Histology is the study of tissues, cytology of cell structure and embryology of development.

Define or describe each of the following:
1. anatomy_____
2. physiology_____
3. histology_____
4. embryology_____
5. pathology_____
6. cytology_____

TEST YOURSELF

Match the following:
1. anatomy __C__ body structure   a. disease processes
2. physiology __E__ body function   b. normal tissue
3. pathology __A__ disease processes   c. body structure
4. histology __B__ normal tissue   d. development
5. embryology __D__ development   e. body function
                                                  f. cell organelles

LO 2: Define and use properly the principal
orientational terms employed in human anatomy.

The principal anatomic directional terms are as follows: Cephalic (superior)--upward. Inferior--downward. Ventral (anterior)---in front. Dorsal (posterior)--behind. Transverse---at right angles to body axis. Sagittal--parallel to body axis, passing through the anterior and posterior surfaces. Lateral--far from the body axis. Medial---near to the body axis. Distal--far from the point of origin of a limb. Proximal--near to the point of origin of a limb.

Indicate by labeling the following diagram what is meant by anterior, posterior, superior, inferior, medial, lateral, proximal and distal.

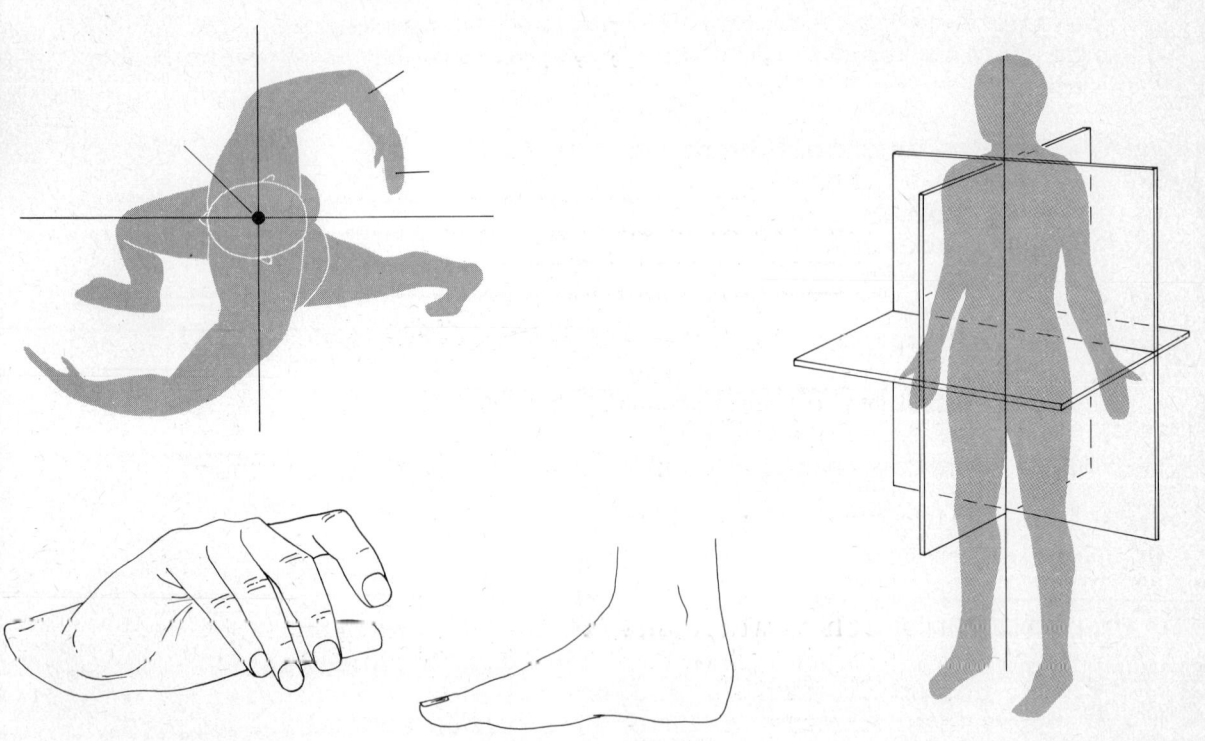

TEST YOURSELF

Match the following:
6. The finger is ____ to the wrist
7. The elbow is ____ to the wrist
8. The head is ____ to the neck
9. The waist is ____ to the crotch
10. The navel is ____ to the intestines

a. superior
b. distal
c. inferior
d. proximal
e. anterior

LO 3: List and locate the principal regions and natural cavities of the human body.

The natural cavities of the human body are classed as a dorsal bony neural cavity and a ventral visceral cavity. The visceral cavity is in turn divided into the chest or thoracic cavity and the abdomino-pelvic cavity. Further subdivisions of the abdomino-pelvic cavity are the abdominal and pelvic cavities, and of the neural cavity are the cranial cavity and vertebral canal.

1. Locate the following:
   (a) axial region_____
   (b) thorax_____
   (c) abdomen_____
   (d) pelvis_____

2. (a) Which two cavities are separated by the diaphragm _____ and _____
   (b) With which <u>major</u> body cavity is the male scrotal cavity associated?_____
   (c) What organ does the cranial cavity contain? _____

<u>TEST</u> <u>YOURSELF</u>

(In this and other such questions, the same answer MAY be used more than once.)
Match the following:
11. urinary bladder_____   a. abdominal cavity
12. stomach_____   b. mediastinum
13. heart_____   c. pelvic cavity
14. testes_____   d. vertebral canal
15. spinal cord_____   e. scrotal cavity

LO 4: Define homeostasis and show that it is a basic functional principle of human physiology, giving examples.

Homeostasis refers to the ability of the body to maintain a constant internal environment at all times which is ideal for the life of its tissues and their proper function. In general, homeostasis is maintained by negative feedback mechanisms, in which a product, when present in excess, inhibits its own production.

1. What is homeostasis? _ability of the body_

2. Give an example of homeostasis <u>other</u> than that provided in the textbook _inhibits its own product_

3. Briefly describe how the body regulates its temperature _____

TEST YOURSELF

16. Homeostasis exists when
    a. a set of conditions are average
    b. a set of conditions are at their maximum
    c. conditions tend to resist change
    d. the internal environment fluctuates widely
    e. when there is too little salt content in the blood

17. When a product inhibits its own formation, we call the condition
    a. negative feedback
    b. positive feedback
    c. poikilostasis
    d. incrementation
    e. hologeny

> *LO 5: Define systems, organs, and tissues and be able to summarize the major body systems in terms of their component organs, function, and contribution to body homeostasis.*

The organism is composed of ten cooperating body systems which are made up of functionally integrated organs. Each organ performs some special function in the body and is made up of specialized tissues, that is, homogenous associations of cells and cell products.
The ten systems are:
1. *integumentary*--retards evaporation; protects against abrasion and infection
2. *muscular*--moves body and individual body organs
3. *urinary*--regulates blood composition; excretes wastes
4. *endocrine*--regulates body function by means of substances secreted into the blood stream
5. *skeletal*--supports body, transmits muscular force, protects organs, stores calcium and phosphate
6. *nervous*--controls most body functions
7. *reproductive*--transmits genetic information
8. *digestive*--chemically simplifies food and transfers it to the blood
9. *circulatory*--carries oxygen and food to cells; removes waste products; provides chemically suitable cellular environment; carries hormones
10. *respiratory*--transfers oxygen into blood; removes CO_2

1. What is a tissue?_____
   _____

2. What is an organ?_____
   _____

3. What is a system?_____

4. In the space below, describe the relationship among the three:

Note: Be sure that you can name all ten body systems
      from memory.

TEST YOURSELF

Match the following:
18. Protects from bacterial   a. skeletal system
    invasion___   b. urinary system
19. Regulates blood composi-   c. integumentary system
    tion___   d. muscular system
20. Charges blood with   e. respiratory system
    oxygen___
21. Supporting framework of
    body___
22. Moves body___

> LO 6: Taking the human thumb as an exemplary
> organ, summarize its structure and function.

The thumb is perhaps the most important of the digits
because it can be opposed to the fingers, making possible the distinctive human grip. The thumb consists
principally of muscle and bone, the muscles being so
arranged as to antagonize one another's actions. The
thumb is extensively padded with connective tissue and
fat, and coated with skin plus the distal, dorsal nail.
Blood flows into the thumb through arteries, and via
minute branches (capillaries) supplies the tissues
with food and oxygen. Spent blood flows back out
through the veins.

1. What is the basic action of all muscles?_____

2. What are tendons?_____
   _____

3. Why do the thumb and fingers need tendons?_____
   _____

4. What are the three main kinds of blood vessel and
   what is the function of each? (a)_____
   (b)_____
   (c)_____

TEST YOURSELF

23. The basic muscle action is to
    a. pull   b. push   c. expand
    d. rotate   e. any of the above

24. Which of the following functions does a capillary perform?
    a. carries blood toward the heart
    b. carries blood away from the heart
    c. brings blood and its components close to the body cells
    d. transmits control impulses

> *LO 7: Summarize the principal characteristics, functions and locations of the main body tissues, giving special notice to the epithelial and connective tissues.*

The body possesses four principal tissues: muscular, nervous, epithelial, and connective. Muscle cells, specialized for contraction, are elongated and voluntary muscle is multinucleate. Nerve cells, specialized for transmission, are even more elongated than muscle cells. Connective tissue is difficult to summarize but is mainly adapted to supportive and mechanical tasks, while epithelium is specialized for protection, secretion, absorption and even movement in some cases.

1. Give the four main tissues and their distinctive characteristics:
   a. _____

   _____

   b. _____

   _____

   c. _____

   d. _____

2. To learn the epithelial and connective tissues, we suggest the use of flash cards. Using table 1-2 of the text as your source, write the name of the tissue on one side of each card plus, if possible, a sketch. On the other, give other pertinent information, especially description and functions. Quiz yourself by consulting one side of a card, trying to recall the material on the other side. Turn the card over if you get stuck, or to check yourself.
3. How is it that the blood may be regarded as a connective tissue?_____
_____
4. Compare bone and cartilage_____
_____
_____
_____

5. In the space below, give the three major types of muscle, their location in the body, and compare their function.

TEST YOURSELF

Match these epithelial tissues:
25. Nuclei at different levels, but all cells contact the same basement membrane ____
26. lines the bladder ____
27. skin ____
28. occurs in the lining of the respiratory tract ____
29. occurs in kidney tubules ____

a. cuboidal
b. stratified squamous
c. transitional stratified
d. psuedostratified
e. simple columnar

Match these connective tissues:
30. energy storage ____
31. fluid tissue ____
32. cells isolated in individual spaces within mineralized, rigid matrix ____
33. Cartilage with no visible fibers ____
34. tendons ____

a. white, fibrous
b. hyaline
c. bone
d. blood
e. adipose

ANSWERS TO ALL TEST-YOURSELF QUESTIONS:

1. c   2. e   3. a   4. b   5. d   6. b   7. d   8. a   9. a
10. e 11. c 12. a 13. b 14. e 15. d 16. c 17. a
18. c 19. b 20. e 21. a 22. d 23. a 24. c 25. d
26. c 27. b 28. e 29. a 30. e 31. d 32. c 33. b
34. a

POST-TEST

1. (LO 1) The study of normal tissue structure is called a. histology b. gross anatomy c. physiology d. pathology e. cytology

2. (LO 2) The tip of the nose is ____ to its root (attachment to the face) a. superior b. inferior c. anterior d. posterior e. none of the above

3. (LO 2) The sole of the foot is called its ____ aspect a. dorsal b. palmar c. plantar d. anterior e. posterior

4. (LO 3) The mediastinum divides the two halves of the ____ cavity   a. abdominopelvic (b.) thoracic   c. cranial   d. scrotal   e. dorsal

5. (LO 3) The ____ is part of the appendicular portion of the body   a. head   b. foot   c. elbow   d. chest   (e.) two of the preceding answers are correct

6. (LO 4) The next level of body organization above the cell is the   (a.) tissue   b. organ   c. system   d. organelle   e. organismic

7. (LO 4) Which of the following body systems might almost be said to consist of a single organ   a. respiratory   b. nervous   c. urinary   (d.) integumentary   e. circulatory

8. (LO 4) Which of the following pairs of systems really comprise a single functional system   (a.) muscular-skeletal   b. nervous-circulatory   c. digestive-skeletal   d. endocrine-urinary   e. reproductive-circulatory

9. (LO 5) Which of the following are NOT located in the thumb   a. veins and arteries   b. capillaries   (c.) extrinsic thumb muscles   d. tendons   e. articular cartilages

10. (LO 6) The type of tissue that lines the digestive tube is   a. connective tissue   b. muscle tissue   c. ER tissue   d. nervous tissue   (e.) epithelial tissue

11. (LO 6) Tissue in which cells are separated from one another by intercellular substance is called   a. epithelial tissue   (b.) connective tissue   c. muscle tissue   d. nervous tissue   e. nuclear tissue

12. (LO 6) ____ occurs in the intervertebral discs   a. hyaline cartilage   (b.) fibrocartilage   c. white fibrous connective tissue   d. yellow elastic connective tissue   e. areolar connective tissue

13. (LO 6) In which of the following epithelial tissues are the outermost cells usually dead   a. simple squamous   b. stratified squamous   c. psuedostratified columnar   d. stratified cuboidal   e. stratified columnar

14. (LO 6) A tissue is best defined as   a. a group of similar cells associated to perform a common function   b. a group of similar cells which unite to form a compact organ   c. a group of cells which unite to form a thin sheet or membrane   d. a group of different kinds of cells that work together to perform a task   e. answers (b) and (c) are both correct

15. (LO 6) The type of tissue that covers or lines body surfaces is called   a. epithelium   b. muscle   c. connective tissue   d. lysosome tissue   e. nervous tissue

16. (LO 6) Connective tissue is specialized to   a. contract   b. support and hold together other tissues   c. cover body surfaces   d. receive and transmit impulses   e. all of the above are true

17. (LO 6) Tissue specialized to receive and transmit impulses is called   a. placental tissue   b. mesoderm   c. connective tissue   d. nervous tissue   e. epithelium

18. (LO 6) Tissue that is specialized for contraction is called   a. epithelium   b. connective tissue   c. nervous tissue   d. ectoderm   e. muscle

NOTE: Essay-type questions covering this material will be found at the end of Chapter one in the textbook.

ANSWERS TO POST TEST QUESTIONS

1.a   2.c   3.c   4.b   5.e   6.a   7.d   8.a   9.c
10.e   11.b   12.b   13.b   14.a   15.a   16.b   17.d
18.e

2 CHEMISTRY AND LIFE

PRETEST

1. An atom consists of two main parts, the _____ and the _____ of which is composed of yet smaller units called _____ particles
2. _____ are negatively charged particles which move in orbits about the nucleus of an atom
3. C is the chemical symbol for _____, an element found in all _____ compounds. Fe is the chemical symbol for _____, which occurs in the _____ of the blood
4. _____ are built from molecular subunits called amino acids
5. Monosaccharides are simple _____
6. _____ fatty acids contain the maximum number of hydrogen ions that is chemically possible
7. The formula CO_2 denotes a molecule which consists of _____ (no.) atoms of _____ and _____ (no.) atoms of _____
8. The main constituents of the cell membrane are _____ and _____
9. The basic arrangement of the cell membrane is _____ layered
10. _____ are organic catalysts that regulate chemical reactions in living organisms
11. The substance that an enzyme acts upon is called a _____
12. The protein part of an enzyme is known as its _____, but a _____ is also a vital part of the enzyme molecule
13. Allosteric control of enzymes depends upon changes in the _____ of the enzyme molecule

14

14. What is wrong with this diagram of the cell membrane?

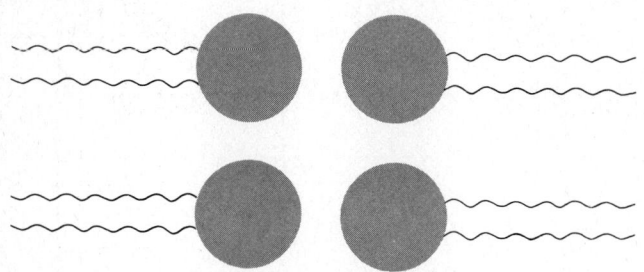

15. The active site(s) of an enzyme form(s) a temporary chemical bond with the _____(s)
16. The atoms of a molecule are held together by forces of attraction between them termed_____,
17. In a _____ bond, atoms share electrons
18. Three types of ionic compounds are_____, _____ and_____
19. When ordinary table salt is dissolved in water, what ions are liberated: _____ and_____
20. When an atom loses electrons by donating them to another atom, the bond thus formed is said to be _____
21. A pH of 10 is_____
22. pH refers to the concentration of _____ in a solution
23. a_____ tends to minimize changes in pH
24. The body controls undesirably low blood pH partly by chemically replacing _____ acids by _____ ones

ANSWERS:
1. nucleus and orbitals; subatomic (LO 1) 2. electrons (LO 1) 3. carbon, organic; iron, hemoglobin (LO 1)
4. proteins (LO 2) 5. sugars (or carbohydrate)(LO 2)
6. saturated (LO 2) 7. one, carbon; two, oxygen (LO2)
8. lipid (or phospholipid) and protein (LO 3)
9. two (LO 3) 10. the hydrophobic ends of the molecules face outwards (LO 3) 11. enzymes (LO 4)
12. substrate (LO 4) 13. apoenzyme; coenzyme (LO 4)

15

14. shape (LO 4)   15. substrate (LO 4)   16. chemical bonds (LO 5)   17. covalent (LO 5)   18. acids, bases (or alkalis), and salts (LO 5)   19. Na^+ and Cl^- (LO 5)
20. electrovalent (LO 5)   21. basic (or alkaline)
22. hydrogen ions (LO 6)   23. buffer (LO 7)
24. strong, weak (in that order) (LO 7)

> LO 1: Describe the basic structure of the atom in accordance with the conventions presented in this unit, showing the position of the main subatomic particles: electrons, protons and neutrons, and be able to give the symbols employed in chemistry for the more biologically significant kinds of atoms.

For biologic purposes we may consider the basic particle of matter to be the atom. It consists of a nucleus containing protons and neutrons, plus one to several electron shells or orbitals. The principal chemical properties of an atom are mostly determined by the makeup of its outermost orbital.

1. The smallest unit of an element that retains its chemical properties and can participate in chemical reactions is an_____
2. However, an atom is in turn composed of subatomic particles known as _____, _____ and_____
3. Protons and neutrons are located in the atomic _____
4. Each subatomic particle is charged as follows protons____ neutrons____ electrons____
5. If the outermost orbital of an atom contains_____ electrons it is stable
6. Different elements combined in definite proportions constitute a _____ The smallest particle of such a substance is called a _____
7. Write the name of each of the following chemical elements in the space provided next to its symbol
   C_____   H_____   O_____
   Na_____   N_____   Ca_____
   P_____   Cl_____   Fe_____
   S_____

16

8. The gas _____ is required for respiration
9. _____ is an important component of the hemoglobin molecule

TEST YOURSELF

1. The smallest unit of any element that retains its chemical identity is known as a(n)   a. atom   b. electron   c. nucleus   d. molecule   c. ion

2. Electrically neutral particles are likely to occur in which of these atomic parts   a. the outermost electron shell or obital   b. the innermost orbital   c. the nucleus   d. throughout the atom (in chemical combination)

Match the following. A choice may be used more than once, but be sure that it is the most appropriate alternative.

3. Fe _____
4. nucleic acids, bones _____
5. present in all organic compounds and the backbone of all such _____
6. gas required for cell respiration _____
7. P _____
8. occurs both in proteins and nucleic acids _____
9. found in hemoglobin _____

a. carbon
b. phosphorous
c. nitrogen
d. iron
e. oxygen

> LO 2: (a) chemically define carbohydrates, lipids, proteins and nucleic acids; give their subunits; be able to recognize each from its formula (given examples), and summarize the role of each in the life of the body or of the cell. (b) interpret simple chemical formulas, including structural formulas.

Carbohydrates are compounds of C, H and O, in which there is approximately twice as much hydrogen as carbon. Proteins are composed of linked amino acids. Fats are composed of fatty acids and glycerol (but there are other lipids). Nucleic acids are composed of a pentose-

phosphate backbone and several nitrogenous bases. Carbohydrates function primarily as fuel substances; proteins operate for the most part as either enzymes or as structural proteins; fats and other lipids serve as energy storage substances and help to make up the cell membrane. Nucleic acids govern the genetic heritage of the organism and its expression.

1. A shorthand means of expressing the composition of a molecule is called a_____ _____
2. A molecule consists of two or more chemically_____ atoms united with one another. It is also defined as the smallest particle of a_____ _____ that retains its characteristic properties.
3. A molecule consisting of two or more different types of atoms combined in definite proportions is called a_____ _____
4. In a chemical formula, subscripts are used to denote the_____ of each kind of atom that is present
5. The structural formula of a compound indicates the way in which its component atoms are_____
6. Molecules which have different arrangements of the same atoms, and the same proportions of those atoms are_____
7. Organic compounds are relatively_____, and their main backbone, or chain, is composed of _____atoms
8. Carbon atoms form strong_____bonds with one another
9. Generally, carbon atoms form _____bonds with the other atoms with which they are associated
10. Large organic molecules made up of chains of similar units are called biopolymers. The units that make up _____are amino acids. The units that make up _____are hexoses. The_____ _____ are also made up of repeating units.
11. A carbohydrate molecule has the shape:

    It is a(n)_____
12. A very large molecule consists of many units held together each by a C-N bond. It is a(n)_____ _____

13. A large, complicated molecule with a phosphate-sugar backbone is probably a(n)_____

TEST YOURSELF

10. Which of the following would be a chemical compound
    a. air b. salt water c. bread dough d. iron chloride, $FeCl_3$ e. pure calcium

11. Glucose and fructose, two sugars, differ chemically but have the same empirical formula, $C_6H_{12}O_6$. They are a. salts b. alkalis c. isomers d. acids e. bases

12. Which of the following is an organic compound
    a. Na_2CO_3 b. H_2SO_4 c. $CaNO_3$ d. C_3H_7OH e. HCl

Match the following
13. Long chain of hexose units_____   a. isomers
14. Long chain of amino acids_____   b. fat
15. Contains glycerol_____   c. nucleic acid
16. sugar, phosphate, and nucleotides   d. protein
    arranged in a long chain_____   e. polysaccharide
17. fructose and glucose, both
    $C_6H_{12}O_6$

> LO 3: Describe the structure and basic functions of the cell membrane, and summarize the relationship between its structure and its function

The cell membrane is a lipid bilayer with embedded protein granules. The membrane may be pierced with pores. It maintains the chemical and physical differences between the exterior and the cellular interior, and it serves as a port of entry into the cell. (The cell membrane is more precisely known as the plasma membrane).

1. Carefully examine the diagram on the next page. Compare it with Fig. 2-4 in the textbook, then label it from memory

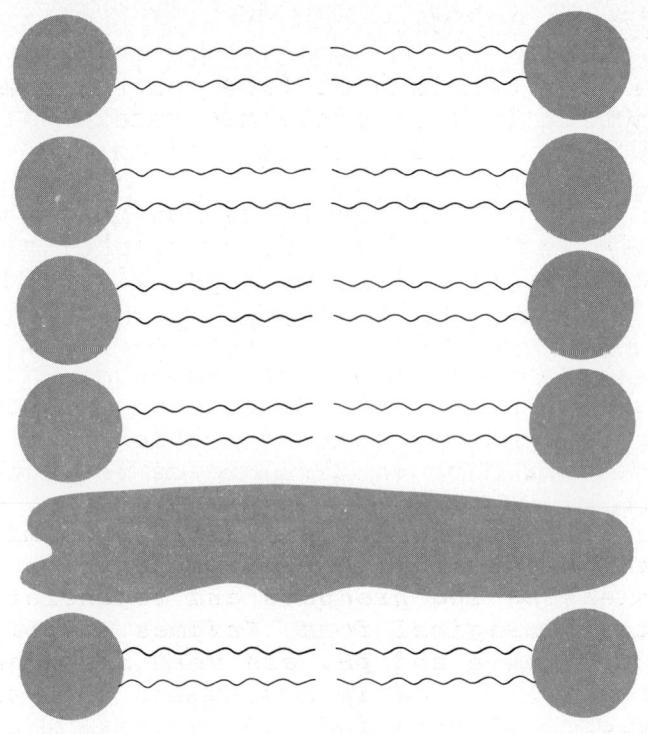

2. What is a fatty acid_____
3. Describe the chemical nature of a typical fat, for example, a triglyceride_____
   _____
4. What is the difference between a monoglyceride, a diglyceride and a triglyceride_____
5. What makes the difference between a completely saturated and a less saturated fatty acid_____
6. Where do phospholipids occur in the living cell _____
   _____
7. (a) What is a lipid bilayer_____
   _____(b) Why does a lipid bilayer form_____
8. What does the function of protein in the cell membrane seem to be_____
   _____

TEST YOURSELF

18. A typical fat consists of from one to three fatty acids chemically joined to a. protein b. a hexose c. glycerol d. phosphate e. starch

19. A fatty acid rich in double carbon-to-carbon linkages is termed a. a wax b. a phospholipid c. unsaturated d. saturated e. a polyglyceride

> LO 4: Describe the function of enzymes, give their principal properties, summarize their chemical makeup, give the enzyme-substrate complex theory of their action, and summarize the main factors that affect their function.

Enzymes function as organic catalysts by forming temporary chemical compounds with their substrates which then break up, releasing the products and regenerating the enzymes in their original form. Enzymes exhibit sharp optima of temperature and pH, are very specific in their substrate preference, and in some cases are controlled by conformational changes induced by certain substances.

Chemically, an enzyme consists of a globular protein molecule combined with another, smaller nonprotein molecule. The components are known as the apoenzyme and the coenzyme, respectively. Both are necessary for the proper function of the enzyme, though the nature of the enzyme depends mostly upon the amino acid sequence of the apoenzyme. Enzymes may be temporarily inhibited or permanently damaged by substances chemically resembling their normal substrates or which damage their active sites.

It is sometimes useful to distinguish between extracellular enzymes and intracellular enzymes. Such enzymes as digestive enzymes and those of the blood plasma or tissue fluid exemplify the extracellular, although it is of course true that they originated as secretory products of cells and therefore originated within them. Intracellular enzymes never leave the cell. The enzymes of cellular respiration afford an example. Vastly more intracellular enzymes are known, for each step in the *complex chemistry of the cell is promoted by a single, specific enzyme.*

1. An enzyme is_____
_____
(Note to student: Please be careful not to identify an enzyme as a digestive chemical. Of course it is true that enzymes occur in the digestive juices, but by far the vast majority of enzymes are intracellular, that is, inside the cell, and have nothing to do with digestion.)
2. Factors that influence enzymatic activity
    a._____
    b._____
    c._____
    d._____
3. What produces the properties of enzymes, or of any proteins_____
_____
4. Why are enzymes important in the life of the cell
_____
_____
_____

5. What are the parts of an enzyme_____
_____
6. What is a catalyst_____
7. What is a substrate_____
8. What do we mean by the "enzyme-substrate complex" and why is it important_____
_____
_____

9. What is an active site_____
_____
10. What is an allosteric site_____

11. How do enzyme inhibitors work_____
_____
Are there more than one kind_____
_____

TEST YOURSELF

20. Enzymes are a. organic catalysts b. designated by the suffix -ase c. generally altered permanently by the chemical reaction in which they participate d. answers (a),(b) and (c) are correct e. answers Only answers (a) and (b) are correct

21. Enzymes may be inactivated by a. very high temperature b. certain inhibitors c. inappropriate pH d. answers (a),(b) and (c) are correct e. only two of the preceding answers are correct

22. A coenzyme is a. part of an enzyme b. necessary for the enzyme to function c. not a protein d. all of the preceding e. (a) and (b) only

23. A substrate is a. the substance produced by an enzyme b. the substance upon which an enzyme acts c. the basic part of the enzyme to which the coenzyme is attached d. a vitamin e. a membrane, such as the cell membrane, to which enzymes or other proteins are attached

24. The shape of an enzyme can often be changed by action of a. the allosteric site b. the active site c. the coenzyme d. the product e. none of the preceding (and in this way, the action of an enzyme may be controlled)

25. Penicillin interferes with bacterial growth by a. interfering with cell wall construction b. enzyme inhibition c. damaging the DNA of the bacterial cell d. poisoning the enzymes used in bacterial respiration e. (a) and (b)

> *LO 5: Define covalence and electrovalence, and relate electrovalent bonding to ionization. Be able to summarize the ionization of common electrovalently bonded substances, especially acids.*

*If an atom tends to lose or gain one or more electrons
from its outermost orbital it is capable of forming
electrovalent bonds with other atoms. Another type of
chemical bond is the covalent bond, which is established
by electron sharing. Only electrovalently bonded substances are able to dissociate in water into the charged
particles called ions. Positively charged (metallic)
ions are called cations. Negatively charged ions are
called anions. Acids, which are electrovalently bonded
compounds of hydrogen, dissociate into hydrogen ions
plus an anion. The strength of an acid is proportional
to the degree to which it dissociates*

1. An atom with more than four electrons in its outer-most orbital tends to_____ more electrons, but if it has less than four, it tends to_____them
2. The number of electrons that an atom can give or accept is referred to as its_____
3. Metals are substances which characteristically _____electrons, while chemically active nonmetals_____them
4. A metal exhibits_____valence, while a nonmetal exhibits _____valence
5. The forces of attraction which hold atoms together into the molecules are termed_____
6. An electrovalent bond is formed when electrons are _____ and _____
7. When a metal atom loses an electron in the estab-lishment of an electrovalent bond, the atom becomes _____charged
8. When an atom gains an electron in similar fashion, it becomes_____charged
9. When an electrovalently bonded compound is dissolved in water, it separates into _____particles called _____
10. Bonds that are established by the sharing of elect-rons are called _____bonds
11. Positively charged ions are called _____ while negatively charged ions are known as_____
12. Complex ions such as $SO_4^=$ consist of groups of atoms _____bonded to one another

24

TEST YOURSELF

26. An atom (other than helium) which has two electrons in its outer shell   a. is a nonmetal   b. will tend to lose two electrons   c. will tend to gain two electrons   d. is likely to either gain or lose electrons depending upon circumstances

27. An atom has 10 protons in its nucleus. It is a. a metal   b. a nonmetal   c. able to behave either as a metal or as a nonmetal   d. probably incapable of entering into any chemical reactions   e. a hydrogen atom

28. When potassium and iodine are allowed to react together, the potassium atoms each lose an electron, and the iodine atoms each gain that electron, so that in the resulting compound, potassium iodide, each atom now possesses an outer electron shell containing <u>eight</u> electrons. In the solid, undissolved state, the resulting molecules may be said to be: a. extremely unstable   b. electrovalently bonded   c. sharing electrons   d. electrically charged   e. acids

29. Which of these transactions establishes a covalent bond   a. sodium donates an electron to chlorine   b. two chlorine atoms mutually share some of the electrons in their outermost shells   c. an organic compound loses hydrogen   d. a chemical reaction produces heat as a by-product   e. chlorine accepts an electron from hydrogen

30. Which of the following consists of covalently bonded atoms
    a. Na^+Cl^- b. K^+I^- c. $Fe^{+++}Cl_3^-$ d. PO_4^{---}
    e. none of the preceding

> LO 6: Define pH and be able to identify any given pH as acidic, basic or neutral.

pH is a measure of acidity or alkalinity that is based upon the concentration of hydrogen ions in a solution. Any pH greater than 7 is alkaline or basic, any pH less than 7 is acidic, and a pH exactly equal to 7 is neutral.

1. An acid is a compound, which when dissolved in water yields_____
2. A base is a compound, which when dissolved in water yields_____
3. A pH value of 7 is_____
   a value of less than 7 is _____
   and a value exceeding 7 is _____
4. The higher the concentration of hydrogen ions in a solution, the _____ its pH will be
5. The strength of an acid is a consequence of the degree to which it _____
6. A pH of 3 is_____ times more/less than a pH of 6

TEST YOURSELF

31. A compound which dissociates in solution to release hydrogen ions is called a. an anion b. a base c. an alkali d. an acid e. a hydroxyl group

32. Which of the following could be the pH of the solution mentioned in question 31 a. 8.5 b. 7 c. 9 d. 6.5 e. 14

33. Which of the following could be expected to possess a pH of greater than 7 a. a hydrochloric acid solution b. a carbonic acid solution c. a solution with fewer hydrogen ions than water d. a solution with a higher concentration of hydrogen ions than water e. none of the above

> LO 7: Describe the way in which buffers minimize changes in pH.

A buffer is usually a solution of a weak acid plus a salt of that acid. The addition of a strong acid forms a salt of that acid, liberating the weak acid. Since the weak acid will have a higher pH than the strong, the change in pH experienced by the solution is less than would be the case if the buffer were not present. Buffers also minimize the effect of alkali upon pH, for the weak acid they contain will combine with the alkali.

1. A buffer pair consists of a weak _____ plus a _____ of that _____
2. In buffering reductions of pH, a _____ acid is replaced with a _____
3. If a strong alkali were to enter the bloodstream, its hydroxyl ions would react with _____ _____

TEST YOURSELF

34. A buffer consists of   a. a weak acid   b. a strong base   c. a salt   c. (a) and (c)   d. (b) and (c)

35. A buffer would minimize the effect of a strong acid by   a. reducing the concentration of hydrogen ions   b. reducing the anion concentration   c. removing the salt of the acid from the solution   d. reducing the pH

ANSWERS TO ALL TEST YOURSELF QUESTIONS

1.a	2.c	3.d	4.b	5.a	6.e	7.b	8.c	9.d
10.d	11.c	12.d	13.e	14.d	15.b	16.c	17.a
18.c	19.c	20.e	21.d	22.d	23.b	24.a	25.e
26.b	27.d	28.b	29.b	30.d	31.d	32.d	33.c
34.c	35.a

POST-TEST

1. (LO 1) The nucleus of an atom contains   a. electrons   b. protons   c. neutrons   d. answers (a),(b) and (c) are correct   e. only answers (b) and (c) are correct

2. (LO 1) The correct chemical symbol for calcium is   a. C   b. Ca   c. Cal   d. Cl   e. Mg

3. (LO 1) The chemical properties of an atom are mostly determined by   a. the neutrons of the nucleus   b. the protons of the first orbital   c. the electrons of the first orbital   d. the electrons of the outermost orbital   e. all of the above

4. (LO 1) Different elements combined in definite proportions constitute   a. a mixture   b. solution   c. chemical compound   d. nucleus   e. colloid

5. (LO 2) The formula $CaCl_2$ indicates a compound that consists of   a. one atom of carbon and two atoms of chlorine   b. one atom of calcium and two molecules of chlorine   c. two molecules of calcium and chlorine   d. one atom of calcium and two atoms of chlorine   e. two molecules of calcium chloride

6. (LO 2) Proteins differ from one another with respect to   a. the type of amino acids that compose them   b. the number of amino acids   c. the arrangement of amino acids   d. all the above   e. two of the above

7. (LO 2) An organic chemical has the formula $C_5H_{10}O_5$ It is (a)   a. pentose   b. hexose   c. heptose   d. polysaccharide   e. glucose

8. (LO 2) A nucleic acid would contain   a. amino acids   b. pentose   c. glycerol   d. fatty acids   e. (a) and (d)

9. (LO 2) A compound containing carbon, hydrogen, oxygen and nitrogen arranged into amino acid subunits would be a   a. fat   b. protein   c. nucleic acid   d. carbohydrate   e. salt

10. (LO 3) Which of the following form a major part of the cell membrane   a. triglycerides   b. phospholipids   c. proteins   d. carbohydrates   e. (b) and (c)

11. (LO 3) In the cell membrane, the hydrophyllic ends of the molecules   a. face toward the cellular interior   b. face toward the cellular exterior   c. face toward one another   d. (a) and (b)   e. (a) and (c)

12. Which of the following is not true of enzymes   a. they are always proteins   b. they always act most efficiently at neutral pH   c. they each react best at a specific optimum temperature   d. they are sensitive to certain poisons   e. they are not permanently altered chemically by the reaction they promote (LO4)

13. (LO 4) The shape of an enzyme molecule   a. is similar for most enzymes   b. permits only certain substrate molecules to come into close contact with it and thus with each other   c. is the basis for naming an enzyme   d. all of the above are true   e. (b) and (c) are true

14. (LO 4) Changes in enzyme function resulting from changes in the shape of the molecule are termed   a. allosteric regulation   b. inhibition   c. optimism   d. coenzymism

15. (LO 4) The substrate   a. is a substance acted upon by an enzyme   b. attaches to the enzyme's active site   c. forms a temporary chemical compound with the enzyme molecule   d. is chemically changed by enzyme action into some product   e. all of the above

16. (LO 5) The atoms of a molecule are held together by   a. chemical bonds   b. radiant energy   c. connecting filaments   d. electron bridges   e. proton spin forces

17. (LO 5) In a covalent bond   a. one atom donates electrons to another atom   b. one atom accepts electrons from another atom   c. two atoms share electrons   d. two atoms share protons   e. neutrons are exchanged

18. (LO 5) A cation is   a. usually a metal   b. positively charged   c. electrovalently bonded when not in solution   d. all of the above

19. (LO 6) A solution with a pH of 8.3 would be a(n)   a. acid   b. base   c. salt   d. neutral solution   e. covalently bonded colloid

20. (LO 7) A buffer   a. minimizes pH change   b. prevents electrons from getting too close to the atomic nucleus   c. is responsibe for the covalent bond.   d. is a weak alkali plus a salt of that alkali   e. (a) and (d)

ANSWERS TO POST TEST QUESTIONS
1. e   2. b   3. d   4. c   5. d   6. d   7. a   8. b   9. b
10. b   11. d   12. b   13. b   14. a   15. e   16. a   17. c
18. d   19. b   20. a

3 THE CELL
Basic Unit of Life

PRETEST

1. The cell is considered to be the basic unit of life because:
   (a) _it is a life it self sufficient life_
   (b) _all living have the smallest as all the body can function_
2. The main task of the lymphocytes is to _defend the body from infection_
3. Organelles are specialized structures within the _cell_
4. Every cell is surrounded by a _cell plasma membrane_
5. Ribosomes function as tiny manufacturing plants where _proteins_ are assembled
6. The _golgi apparatus_ functions as a packaging plant within the cell and also produces sacs of digestive enzymes called _lysosome_
7. The power plants of the cell, in which much of the energy used by the cell is liberated are called _mitochondria_
8. When its _size_ approaches is the limits of efficiency, a cell divides to form two new cells.
9. Mitosis provides for the precise duplication and distribution of _chromosomes_
10. During prophase, the nuclear membrane _disapears_
11. Chromosomes are duplicated during _____ (phase)
12. Ordinarily the lymphocyte moves in _amebiod_ fashion. It may sometimes pass between capillary wall cells in a process known as _diapesis_

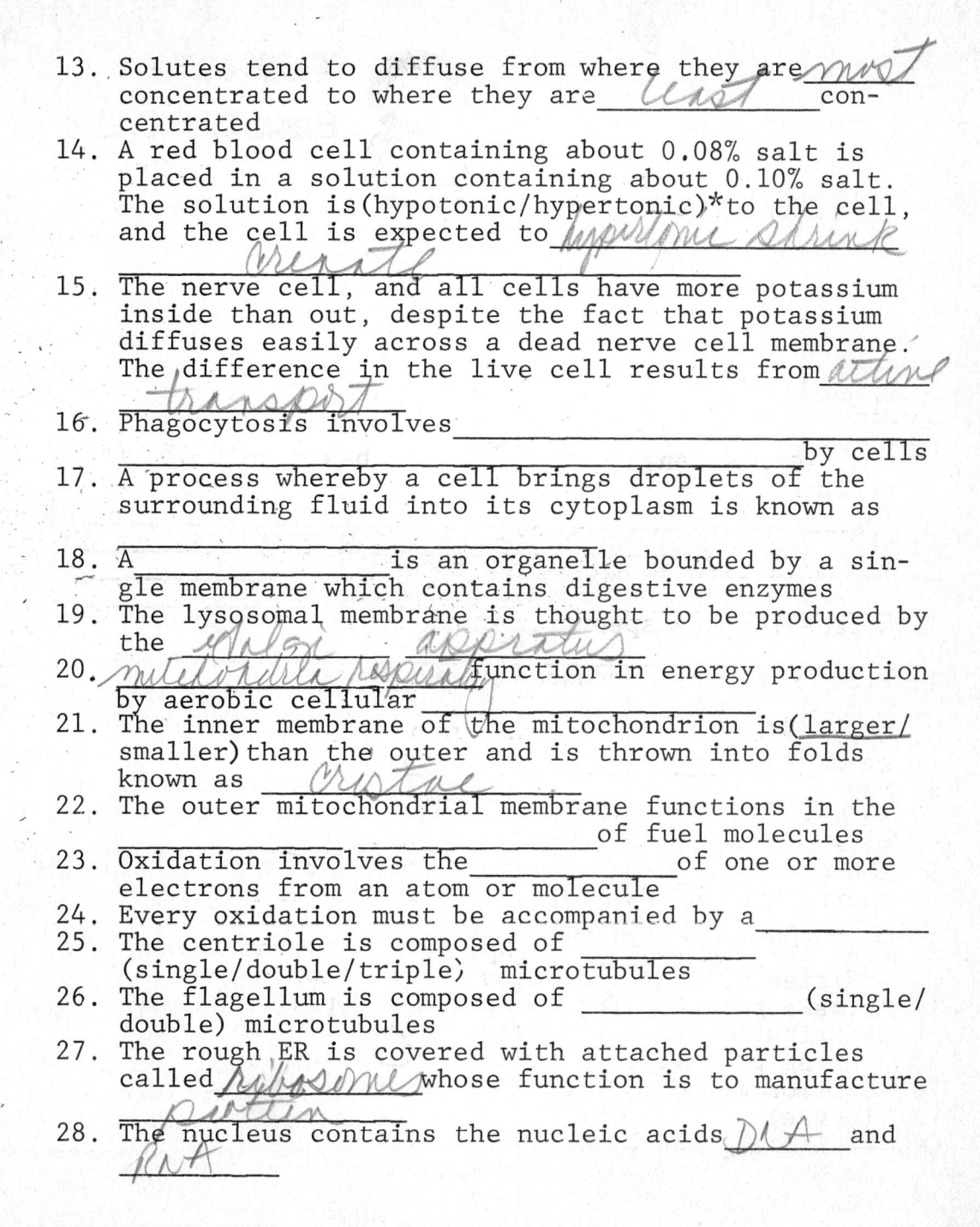

13. Solutes tend to diffuse from where they are _most_ concentrated to where they are _least_ concentrated
14. A red blood cell containing about 0.08% salt is placed in a solution containing about 0.10% salt. The solution is (hypotonic/~~hypertonic~~)* to the cell, and the cell is expected to _hypertonic shrink / crenate_
15. The nerve cell, and all cells have more potassium inside than out, despite the fact that potassium diffuses easily across a dead nerve cell membrane. The difference in the live cell results from _active transport_
16. Phagocytosis involves _____ by cells
17. A process whereby a cell brings droplets of the surrounding fluid into its cytoplasm is known as _____
18. A _____ is an organelle bounded by a single membrane which contains digestive enzymes
19. The lysosomal membrane is thought to be produced by the _golgi apparatus_
20. _mitochondria respire_ function in energy production by aerobic cellular _____
21. The inner membrane of the mitochondrion is (<u>larger</u>/smaller) than the outer and is thrown into folds known as _cristae_
22. The outer mitochondrial membrane functions in the _____ of fuel molecules
23. Oxidation involves the _____ of one or more electrons from an atom or molecule
24. Every oxidation must be accompanied by a _____
25. The centriole is composed of _____ (single/double/triple) microtubules
26. The flagellum is composed of _____ (single/double) microtubules
27. The rough ER is covered with attached particles called _ribosomes_ whose function is to manufacture _protein_
28. The nucleus contains the nucleic acids _DNA_ and _RNA_

_____
*Cross out the inappropriate choices

29. The _____ _____ of the cell is deposited in the nucleus
30. The nucleolus is located within the_____
31. The nucleus is surrounded by a_____

ANSWERS TO PRETEST

1. a) it is capable of self-sufficient life and is a complete metabolic unit   b) all living things are composed of cells (LO 1)   2. defend the body from infection (LO 2)   3. cell (LO 2)   4. cell (plasma) membrane (LO 2)   5. proteins (LO 2, LO 10)   6. Golgi apparatus--lysosomes (LO 2, LO 10)   7. mitochondria (LO2,LO10)   8. size (LO 10)   9. Chromosomes (or genetic information) (LO 12)   10. disappears (LO 12)   11. interphase (LO 12)   12. ameboid(diapedesis) (LO 3) 13. most-least (in that order) (LO 3)   14. hypertonic-shrink (crenate) (LO 4)   15. active transport (LO 5)   16. engulfment of material (LO 6)   17. pinocytosis (LO 6)   18. lysosome (LO 7)   19. Golgi apparatus (LO 7)   20. mitochondria-respiration (LO 8)   21. larger-cristae (LO 8)   22. aerobic-oxidation (LO 8)   23. loss (LO 9)   24. reduction (LO9)   25. **triple** (LO 10) 26. **two--single--nine** (LO 9)   27. ribosomes-protein (LO 10)   28. DNA-RNA (LO 11)   29. hereditary information (LO 11)   30. nucleus (LO 11)   31. membrane (LO 11)

   LO 1: Give two reasons why the cell is considered the basic unit of life.

The cell is considered the basic unit of life (1) because it is the smallest self-sufficient or complete unit in the body, (2) and all true organisms are composed of cells

1._____

2._____

TEST YOURSELF:

1. The smallest unit of living material capable of carrying on all the activities necessary for life is the
   a. tissue b. organ c. nucleus d. atom e. (cell)

2. Which of these is a complete metabolic unit
   a. lysosome   b. cell nucleus   c. mitochondrion
   d. ribosome   e. none of the above

> LO 2: Describe the structure and function of the lymphocyte in general terms and be able to use it as an example of a cell.
> LO 3: Describe lymphocyte locomotion.

The lymphocyte is a white blood cell that attacks bacteria invading the body. The lymphocyte moves by ameboid motion in which the cellular contents flow into psuedopods. A specialized form of this motion is diapedesis, in which the lymphocyte penetrates between adjacent other cells.

1. What is the role of the lymphocyte in the body_____

2. Describe lymphoctye locomotion_____

3. Describe the usual shape of a lymphocyte_____

TEST YOURSELF

3. The principal function of the lymphocyte is in/to
   a. blood transport of oxygen   b. development of disease immunity   c. line blood vessels   d. lay down bone and cartilage   e. transmission of control impulses

4. When a cell squeezes between other cells we call the process   a. dialysis   b. pronation   c. diapedesis   d. pincytosis   e. phagocytosis

> LO 4: Describe the diffusion of the components of liquids, predict the primary direction of diffusion, and relate it to the phenomenon of osmosis.

Both solutes and solvents tend to diffuse from where they are most concentrated in a solution to where they are least concentrated. In osmosis a semipermeable membrane interferes with the diffusion of a solute, but not its solvent (at least not to the same extent). The solvent therefore tends to move toward the hypertonic side, i.e., the side in which the <u>solute</u> is more concentrated. <u>Note</u>: High solute concentration is also called high <u>osmotic pressure</u>; low solute concentration then is low osmotic pressure.

1. Probability favors the diffusion of molecules in a gas or liquid from where they are _____ concentrated to where they are _____ concentrated
2. In osmosis, the _____ passes freely through a differentially permeable membrane, but _____ may not diffuse through it
3. When separated by such a semipermeable membrane, solvent will tend to flow _____ the solution with the higher concentration of solute
4. When the solute concentrations of two solutions are compared, we say that those which are osmotically equal are _____ to one another
5. A solution of relatively high solute concentration is a _____ solution; whereas one of low solute concentration is _____
6. If a solution of 0.85% NaCl is separated from distilled water by a differentially permeable membrane, water flows to the _____
7. If cells are placed in a hypertonic solution, water tends to flow _____ _____ them and they _____

TEST YOURSELF

5. The solute concentration of human blood is equivalent to that of 0.85% sodium chloride. Human red blood cells are placed in a solution of lower solute concentration. They can be expected to a. swell b. shrink c. remain unchanged d. increase their osmotic pressure

6. Which of the following salt solutions is hypotonic to the red blood cell   a. 0.70%   b. 1.50%   c. 3.3%   d. 8.5%

> LO 5: Describe active transport, be able to recognize it when given examples, summarize the piggy-back hypothesis of its mechanism and relate this to the life of the cell.

It is common for differences in solute concentration to exist between the cellular interior and its surrounding medium. In order to produce and maintain these differences cells must actively transport materials across their cell membranes contrary to ordinary tendencies to diffuse. This process, known as active transport, consumes energy. A popular explanation is that a carrier protein in the cell membrane carries the transported substance by means of a conformational change. Active transport is also important in the function of many tissues and organs, for example, in the kidney, and in certain organelles for example, the mitochondria.

1. A process which dynamically opposes diffusion in living cells is called_____ _____
2. Active transport reverses the usual situation of diffusion in that it causes a substance to move from where it is_____concentrated to where it is_____concentrated
3. According to the piggyback hypothesis, special_____ molecules in the cell membrane transport such substances as sodium and potassium
4. The process of active transport consumes_____

TEST YOURSELF

7. Active transport involves: a. concentrating some substance   b. conformational change in some cellular protein   c. consumption of energy   d. all of the preceding

8. Which of these is an example of active transport   a. diffusion of salt into distilled water   b. entry of water into a cell by osmosis   c. crenation   d. the sodium pump   e. the digestion of food

LO 6: Describe pinocytosis and phagocytosis, and summarize the role played by vacuoles in these processes.

Pinocytosis and phagocytosis are accomplished by in-pocketings of the cell membrane which eventually form fluid-filled cavities or vacuoles in the cytoplasm lined with cell membrane. Such vacuoles may function in digestion and/or absorption of materials taken in.

1. Give one type of human body cell that employs phago-cytosis_____ why does it do so_____
   _____
2. What is pinocytosis_____
   How does it differ from phagocytosis_____
   _____

TEST YOURSELF

9. A white blood cell engulfs and digests bacteria by means of   a. pinocytosis   b. phagocytosis   c. diuresis   d. osmosis   e. halitosis

   LO 7: Summarize the nature and digestive function of lysosomes

Lysosomes (vesicles bounded by a single membrane and containing powerful enzymes) function in cellular digestion, development and cell death.

1. What are lysosomes_____
   _____
2. How do lysosomes function in the development of the hand_____
   _____

TEST YOURSELF

10. Digestive enzymes destroy phagocytized bacteria. These enzymes are stored in tiny sacs bounded with a single membrane. These sacs are probably   a. chromosomes   b. centrioles   c. lysosomes   d. desmosomes   e. mitochondria

11. By their action in development, lysosomes: a. kill the cells of the tissue between the fingers b. cause the two halves of the palate to fuse c. produce the early eyes d. cause the early limbs to bud from the trunk d. none of the preceding

> LO 8: Define cellular respiration and describe the structure and function of mitochondria as cellular respiratory organelles.

Mitochondria are double-walled organelles that function in aerobic cellular respiration by the oxidation of organic fuels, particularly glucose. The mitochondrial inner membrane is elaborately folded, is larger than the outer membrane, and bears the enzymes employed in mitochondrial respiration. The outer membrane actively transports fuel materials inside the mitochondrion.

1. What is cellular respiration_____
   _____

2. The mitochondrion consists of_____ membranes. The outer one is involved in_____ of fuel into the organelle, and the inner, (larger/smaller) one does the actual_____

TEST YOURSELF

12. The inner mitochondrial member is a. where respiration occurs b. thrown into folds called <u>cristae</u> c. larger than the outer membrane d. all of the preceding e. (b) and (c) only

13. Which of the following engages in active transport a. mitochondrial outer membrane b. mitochondrial inner membrane d. cristae d. ribosome e. lysosome

> LO 9: Define oxidation in terms of hydrogen or electron loss and relate it to cellular respiration

One definition of oxidation is hydrogen loss. In aerobic respiration organic fuels are initially oxidized in this fashion, releasing some energy. Further energy is released by the ultimate acceptance of hydrogen by oxygen. This forms water. The release of energy by cells is called respiration. Oxidation is also defined as electron loss. Reduction is the opposite of oxidation. The two complementary processes always occur <u>together</u>.

1. Cellular respiration always proceeds in a number of _____
2. During cellular respiration _____ is removed from fuel compounds, and is eventually accepted by oxygen
3. Oxidation may be generally defined as the _____ of electrons, or in the case of covalently bonded compounds, of_____
4. Why must every oxidation be accompanied by a reduction, and vice versa_____
   _____
   _____

TEST YOURSELF

14. In which of the following transaction is iron <u>NOT OXIDIZED</u>   a. $2Fe + 3Cl_2 \longrightarrow 2FeCl_3$
    b. $2FeCl_2 + Cl_2 \longrightarrow 2 FeCl_3$
    c. $2FeI_3 + Cl_2 \longrightarrow 2 FeCl_3 + 6I$
    d. $Fe + 3I \longrightarrow FeI_3$

15. An unsaturated fat would be reduced if a. it were combined with chlorine b. it were converted to a saturated fat c. it were burned d. an electric current were applied to it e. none of the preceding

16. The substance antimony can be treated with chlorine to form the compound antimony trichloride. The chlorine accepts electrons from the antimony. In this instance, antimony behaves as a(n) a. ion b. metal c. nonmetal d. oxidizing agent e. organic compound

17. Which of these is oxidation   a. electron gain   b. hydrogen gain   c. what happens to sodium when it unites with chlorine   d. sugar is dissolved in water   e. two oxygen atoms form a covalent bond

18. In the chemical reaction $2H_2 + O_2 \longrightarrow 2H_2O +$ energy, which is chemically <u>reduced</u>   a. hydrogen   b. oxygen   c. H_2O d. none of the preceding

> LO 10: Describe the structure and summarize the function of (a) centrioles (b) endoplasmic reticulum (c) Golgi apparatus (d) microtubules (e) cilia and flagella

Centrioles are tiny double cylinders composed of hollow tubules that function in mitosis. The endoplasmic reticulum functions in protein synthesis and probably in intracellular transportation. The Golgi apparatus packages cellular secretions and seems to be the source of lysosomal membranes.
Cilia and flagella, hairlike organelles composed of microtubules, function in the propulsion of some cells.

1. Completely describe the structure of a typical microtubule_____

2. What organelles are made up of microtubules_____
_____

3. Diagram a cillium cross-section in the space below

4. A cilium or flagellum consists of 2 single central microtubules surrounded by_____double microtubules
5. What appears to be the function of the arms (or crossbridges) of the ciliary outer microtubules_____

6. How is a centriole different in structure from cilia and flagella_____

7. What is the ER_____

39

8. Describe the rough ER _____
   What does it do _____
9. What is the Golgi apparatus _____
10. Give two functions of the Golgi apparatus
    (a) _____
    (b) _____
11. In what kind of cell is the Golgi apparatus likely to be the most prominent _____

TEST YOURSELF

19. Centrioles consist of a cylindrical arrangement of _____ microtubules which function in the process of _____ a. double--cell locomotion b. single---respiration c. triple--cell division d. circular--active transport e. solid--muscular contraction

20. The smooth ER is composed of a. a system of membranes b. ribosomes c. mitochondria d. microtubules d. (a) and (b)

21. The Golgi apparatus a. is now known to be an artefact, that is, an apparent structure produced by staining procedures which is not actually present in the cell b. packages cellular secretions c. produces the lysosomal membrane d. is responsible for a cellular respiration e. (b) and (c)

22. In cilia and flagella, movement seems to be produced by action of the _____ of the _____
    a. arms--outer microtubules b. ATP---membranous sheath c. contraction--the single, central microtubules d. lengthening--double microtubules e. the mechanism of ciliary action is completely unknown

23. The Golgi apparatus is likely to be most prominent in _____ cells a. muscle b. squamous epithelium c. red blood d. secretory glandular e. there is no rule

LO 11: Describe the structure and chemical makeup of the cell nucleus and of the nucleolus.

The nucleus of the cell is bounded by a porous double membrane and contains DNA, RNA and histone-protein. The nucleolus is basically a cavity within the nucleus that contains RNA for the most part. Most of this RNA is probably associated with ribosome production.

LABEL THIS DIAGRAM:

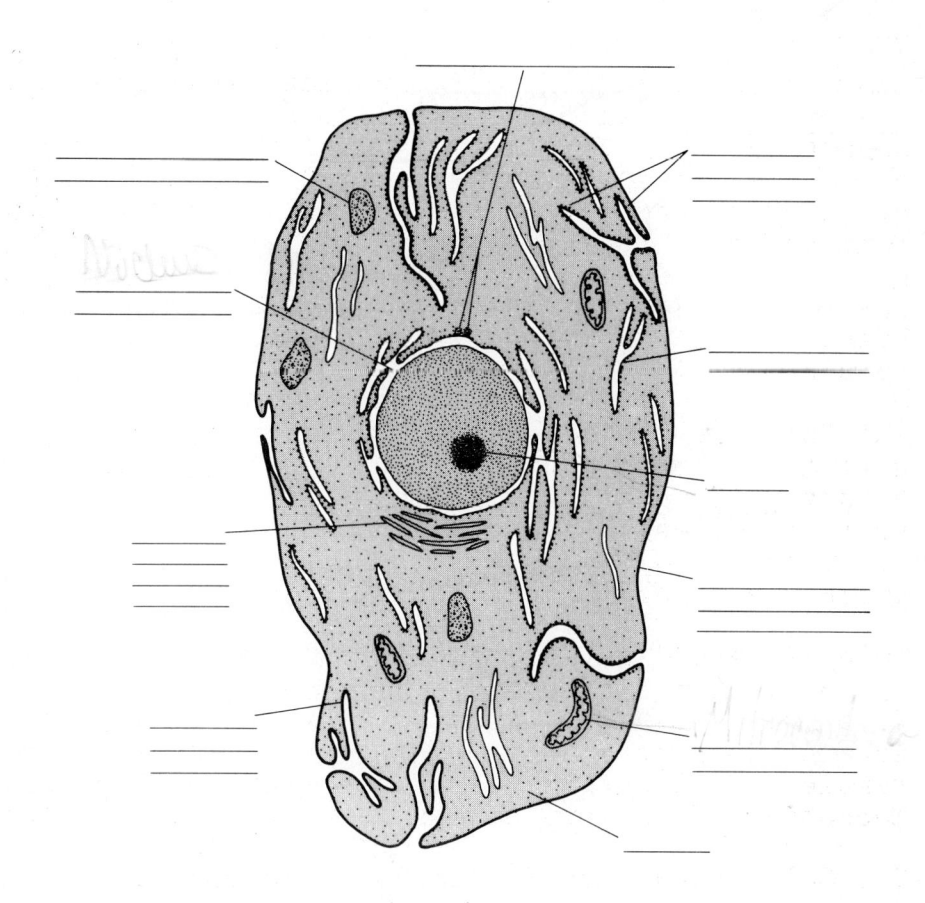

1. The nucleus is surrounded by a porous double,_____
_____

2. The nucleus contains these materials: (1)_____
   (2)_____ (3)_____
3. The nucleolus may produce_____

TEST YOURSELF

24. The genetic code is registered in the   a. RNA
    b. histones   c. proteins   d. phospholipids   e. DNA

25. Except for a small amount in the mitochondria, DNA occurs within the   a. lysosomes   b. nucleus   c. ribosomes   d. rough ER   e. (b) and (d)

> LO 12: Describe the process of mitosis and summarize its functional significance.

SEE TABLE ON FOLLOWING PAGES 44, 45, 46

1. How does mitosis insure chromosome constancy_____
   _____
2. What do spindle fibers do_____
3. What is the function of centromeres_____
4. List the stages of mitosis and the main events of each (1)_____
   (2)_____
   (3)_____
   (4)_____
   (5)_____

NOTE: It may be helpful to remember the acronym IPMAT whose letters stand for each of the five stages of mitosis.

TEST YOURSELF

26. Mitosis always results in   a. four daughter cells, each with half the number of chromosomes as the mother cell   b. two daughter cells, each with half the number of chromosomes as the original cell   c. two daughter cells, each with the identical types and number of chromosomes as the original cell   d. four daugter cells, each with the identical number and types of chromosomes as the

42

original cell e. two daughter cells, each containing the same number of chromosomes, but different types of genes

MATCH THE FOLLOWING

27. Chromatids at the equator of the cell____
28. Cell spends most of its life in this phase____
29. Daughter cells formed by cell cleavage____
30. Nucleolus becomes disorganized

a. interphase
b. prophase
c. telophase
d. anaphase
e. metaphase

ANSWERS TO ALL TEST YOURSELF QUESTIONS

1.e (LO 1)   2.e (LO 1)   3.b (LO 2,3)   4.c (LO 2,3)
5.a (LO 4)   6.a (LO 4)   7.d (LO 5)   8.d (LO 5)
9.b (LO 6)   10.c (LO 7)   11.a (LO 7)   12.d (LO 8)
13.a (LO 8)   14.c (LO 9)   15.b (LO 9)   16.b (LO 9)
17.c (LO 9)   18.b (LO 9)   19.c (LO 10)   20.a (LO 10)
21.e (LO 10)   22.a (LO 10)   23.d (LO 10)   24.e (LO 11)
25.b (LO 11)   26.c (LO 12)   27.e (LO 12)   28.a (LO 12)
29.c (LO 12)   30.b (LO 12)

POST TEST

1. (LO 1) Organs are to organisms as   a. cells:tissues   b. systems:tissues   c. tissues:organs   d. organelles:cells   e. DNA:RNA

2. (LO 1) Which of the following probably could be maintained alive outside the body, under the proper conditions, for quite some time   a. the liver   b. white blood cells   c. connective tissue   d. mitochondria   e. (a),(b) and (c) only

3. (LO 2) The lymphocyte is one of the ____ cells of the body   a. respiratory   b. defensive   c. nervous   d. muscular   e. excretory

4. (LO 3) The lymphocyte's principal means of locomotion is by   a. ameboid movement and diapedesis   b. ciliary movement   c. flagellar locomotion   d. contraction of muscular fibers   e. brachiation

STAGE	DESCRIPTION	COMMENTS
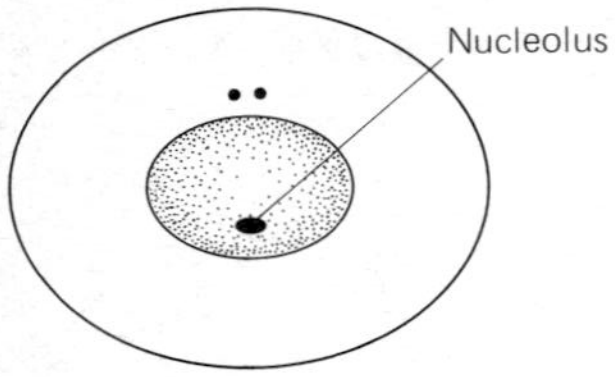 1. INTERPHASE	The normal life of the cell, often called the "resting" stage of the cell. But all life processes except actual division occur in this stage. Toward the end of this phase, the chromosomes duplicate and DNA replicates. It is during interphase that DNA actively produces mRNA, resulting in the formation of cellular protein.	One could make the case that this stage should not be considered a part of cell division as such. Yet if a cell is to divide, by the end of interphase the chromosomes must have duplicated and the DNA must also have been replicated so that both daughter cells will receive a complete copy of the parent cell's genetic information.
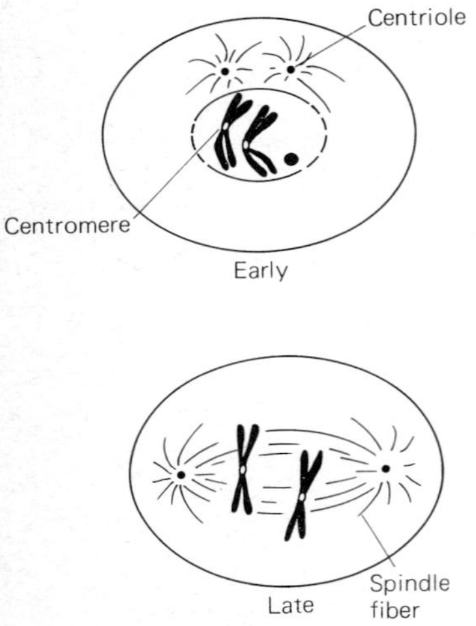 2. PROPHASE	Initial stage of cell division proper. Nuclear membrane disappears, nucleolus becomes disorganized. The chromosomes become apparent, the spindle forms, and the doubled centrioles move toward the poles.	The nuclear structures appear to dissolve into the cytoplasm -- in part, so that the chromosomes can leave the nucleus. The centriole must approach the poles so that, evidently, they may serve as points of attachment and anchorage for spindle fibers. The spindle fibers themselves will serve to pull the chromosomes toward the poles. The chromosomes are mostly DNA and DNA is impractical to transport in its loose, interphase form.

Interphase diagram labels: Nucleolus

Early prophase diagram labels: Centromere, Centriole

Late prophase diagram labels: Spindle fiber

44

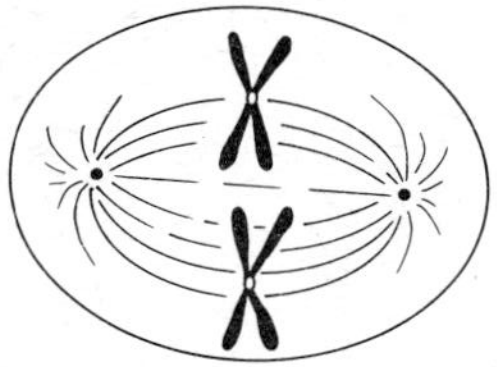

3. METAPHASE

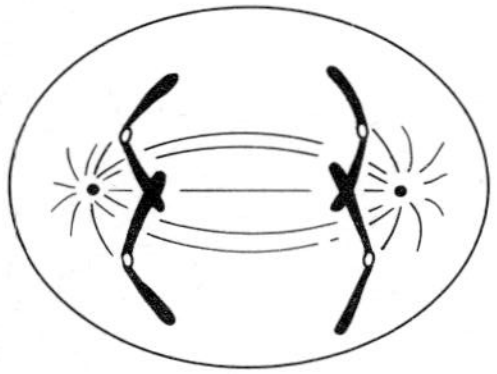

The chromosomes line up along the equator. Actually, they are *double* chromosomes at this stage. Such a double chromosome consists of two chromatids, or single chromosomes. The two are held together by a common central structure, the centromere.

Chromatids are paired, reflecting the replication of genetic information that has already taken place. The members of a chromatid pair are genetic duplicates of each other. The centromeres will serve as moveable points of anchorage for the spindle fibers; the centrioles as immoveable.

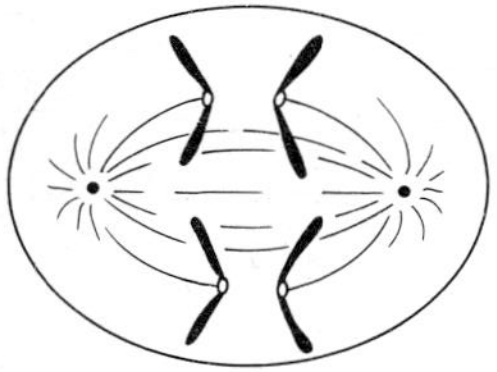

4. ANAPHASE

Centromeres divide, and the chromatids become independent chromosomes. The duplicate chromosomes begin to move toward the opposite poles, drawn by spindle fibers that appear to contract but which are actually probably drawn past one another.

The centromeres must divide in order to allow the chromatids to separate. The chromosomes must now be actually transported to the poles in preparation for the actual physical division of the cell into two daughter cells.

45

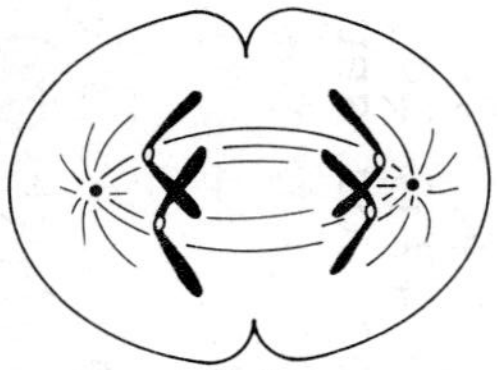

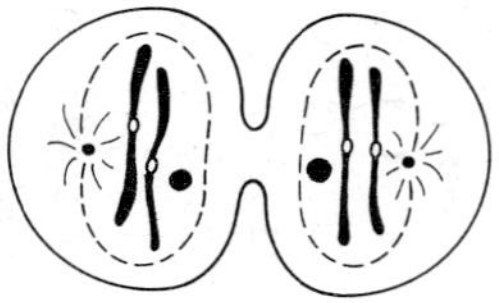

5. TELOPHASE

Cytokinesis, the physical separation of the daughter cells, marks this stage off from late anaphase. Most of the events of prophase now are reversed. The nuclear membrane now reforms, the nucleolus reappears, the chromosomes lose their visual identity and spindle fibers disappear. Centrioles duplicate.

The newly formed daughter cells are now restored to an interphase condition so that they can once again carry out their normal activities. Note that each daughter cell must yet again replicate its DNA if it is to again divide. It is possible that new ribosomes are formed anywhere from early telophase through all of interphase. Their RNA must be made by nuclear DNA but it is hard to see how they could leave the nucleus at times when its membrane is intact.

46

5. (LO 4) In a salt solution, the salt is the   a. solvent   b. colloid   c. solute   d. acid   e. base

6. (LO 4) Some living cells are placed in a hypertonic solution. What happens (assume that that the cell has no sodium pump or other active transport system that can influence the outcome)   a. water flows into the cell   b. the cell swells   c. water flows out of the cell   d. solutes move freely from the cell into the surrounding medium, but water does not move in either direction   e. two of the preceding answers are correct

7. (LO 5) Suppose the cell of No. 6 were to remain unchanged. This result could be explained by   a. osmosis   b. dialysis   c. crenation   d. plasmolysis   e. active transport

8. (LO 6,7) When a white blood cell ingests a bacterium   a. mitochondria attack the bacteria   b. a centriole pushes the bacterium from the cell   c. the nucleus engulfs the germ   d. a lysosome secretes its digestive enzymes onto the bacterium   e. ribosomes cover the germ

9. (LO 7) Lysosomes   a. are filled with digestive enzymes   b. have a single membrane   c. are produced, at least in part, by the Golgi apparatus   d. destroy some cells during development   e. all of the preceding

10. (LO 8) Enzymes needed for cellular respiration are found   a. along the cristae of mitochondria   b. among the granules of the nucleus   c. in the secretions of lysosomes   d. in the Golgi apparatus   e. in the ribosomes of certain (but not all) cells

11. (LO 9) A chemical reaction in which one atom gains electrons while another simultaneously loses them may be termed a(n)   a. ionic exchange   b. electron shelling   c. oxidation-reduction reaction   d. organic reaction   e. transport reaction

The following (numbers 12-18) refer to LO 10

12. composed of triple microtubules ____
13. prominent in secretory cells ____
14. contributes lysosomal membrane ____
15. granular appearance due to ribosomes; manufactures protein ____
16. cellular locomotion ____
17. associated with mitotic apparatus ____
18. 9 double microtubules: 2 single microtubules ____

a. centriole
b. Golgi apparatus
c. rough ER
d. cilia, flagella
e. smooth ER

19. (LO 11) The nucleus is a. the powerhouse of the cell b. a place where ribosomes produce vital protein c. surrounded by a single membrane produced by the mitochondria d. the central library of cellular information e. filled with digestive enzymes

20. (LO 11) Which of the following is NOT surrounded by at least one membrane a. cytoplasm b. mitochondrion c. lysosome d. nucleus e. nucleolus

21. (LO 12) The process by which most cells distribute their chromosomes and divide is called a. phases b. metaphase c. cilia d. interphase e. mitosis

22. (LO 12) During metaphase a. cleavage of the cell takes place b. chromosomes duplicate themselves c. chromatids are distributed along the equator of the cell d. the nuclear membrane reorganizes e. centrioles begin to move apart from one another

23. (LO 12) The centromere a. is a specialized part of the chromosome b. produces spindle fibers c. is composed of triple microtubules d. is sometimes associated with flagella e. forms a part of the ribosome

ANSWERS TO POST TEST QUESTIONS

1.d	2.e	3.b	4.a	5.c	6.c	7.e	8.d	9.e
10.a	11.c	12.a	13.b	14.b	15.c	16.d	17.a	
18.d	19.d	20.e	21.e	22.c	23.a			

4 THE INTEGUMENTARY SYSTEM

PRETEST

1. The integumentary system consists of the _skin_ with its hair, _glands_, _nails_, and other structures
2. Three functions of the integumentary system are _to prevent entry of disease_ and _prevents water loss_ _helps inform central nervous system_
3. The reproductive layer of the epidermis is called _stratum basale_
4. The outermost, horny layer of the epidermis is the _stratum corneum_
5. The layer of skin beneath the epidermis is called the _dermis_
6. The tissue that attaches the skin to the underlying muscle is called _subcutaneous tissue_
7. When a basal cell moves out of stratum basale it enters stratum _spinosum_ and is called a _prickle cell_
8. The tough waterproofing protein of epidermal cells is called _keratin_
9. Dermis is connective tissue composed mainly of _collagen_ fibers
10. Three structures found within the dermis but not within the epidermis are _nerves_ and _blood vessels_ and _hair follicles_
11. The subcutaneous layer consists of connective tissue mainly _adipose_ tissue
12. The root of a hair together with its epithelial and connective tissue coverings is called the _hair follicle_
13. Sebaceous gland secrete an oily substance called _sebum_

49

14. When body temperature increases above normal the activity of the sweat glands __INCREASES__
(increases, decreases)
15. Emotional stress or sexual stimulation increases secretion of the __apocrine__ sweat glands
16. Nails, which are composed mainly of __keratin__, are outgrowths of the __epidermis__
17. Melanocytes inject __pigment granules__ into other epidermal cells
18. Sunburn occurs when the melanin in the skin is unable to absorb all of the __ultraviolet rays__ to which it is exposed
19. Pale skin may indicate __anemia__ while yellowish skin may indicate malfunction of the __liver__
20. Inflammation is characterized by __redness__, __swelling__, __heat__ and __pain__
21. Formation of scar tissue is a process of __collagen__ synthesis and breakdown
22. In __second__ degree burns many epidermal cells die but sufficient numbers remain to repair the wound
23. A __malignant__ neoplasm is one in which the cells grow rapidly and invade surrounding tissues
24. The most common type of skin cancer is __basal cell__ carcinoma
25. Skin cancer is most frequently linked to excessive, chronic exposure to the __sun (ultraviolet rays)__

ANSWERS

1. skin, glands, nails (LO 1) 2. prevents entrance of disease organisms, chemicals, water; prevents water loss helps maintain body temperatures; its sensory receptors help inform central nervous system of changes in environment; excretes excess water and wastes; produces vitamin D. (LO 2) 3. stratum basale (LO 3) 4. stratum corneum (LO 3) 5. dermis (LO 3) 6. subcutaneous tissue (LO 3 or 6) 7. spinosum, prickle cell (LO 4)
8. keratin (LO 4) 9. collagen (LO 5) 10. nerves, blood vessels, glands, hair follicles (LO 5) 11. adipose (LO 6) 12. hair follicle (LO 7) 13. sebum (LO 8)
14. increases (LO 9) 15. apocrine (LO 10) 16. keratin, epidermis (LO 11) 17. pigment granules (LO 12)
18. ultraviolet rays (LO 13) 19. anemia, liver (LO 14)
20. redness, swelling, heat, pain (LO 15) 21. collagen (LO 16) 22. second (LO 17) 23. malignant (LO 18)
24. basal cell (LO 19) 25. sun (ultraviolet rays (LO 19)

LO 1: List the structures that make up the integumentary system.

The integumentary system consists of the skin and the hair, nails and glands.

Why do you suppose that hair, nails, and glands are considered part of the integumentary system_____
_____

TEST YOURSELF

1. Which of the following is not considered part of the integumentary system   a. nails   b. muscle   c. glands   d. hair   e. skin

> LO 2: List 7 functions of the integumentary system and tell how each is homeostatic to the organism.

The integumentary system functions to: (1) prevent water loss, (2) prevent entrance of water from the external environment, (3) prevent entrance of disease organisms and of many harmful chemicals, (4) help maintain constant body temperature, (5) help inform the central nervous system of changes in the external environment via its sensory receptors, (6) help in excretion of excess water and body wastes, (7) produce vitamin D.

1. How would homeostasis be affected if the sweat glands did not function_____

2. How does the skin enable us to swim without upsetting the internal balance of the body_____
_____

TEST YOURSELF

2. Which of the following might be upset if the skin were not functioning   a. water balance   b. body temperature   c. vitamin D balance   d. flow of information to the central nervous system   e. all of the preceding answers are correct

51

LO 3: Draw and label the diagram of the skin including the strata of the epidermis.

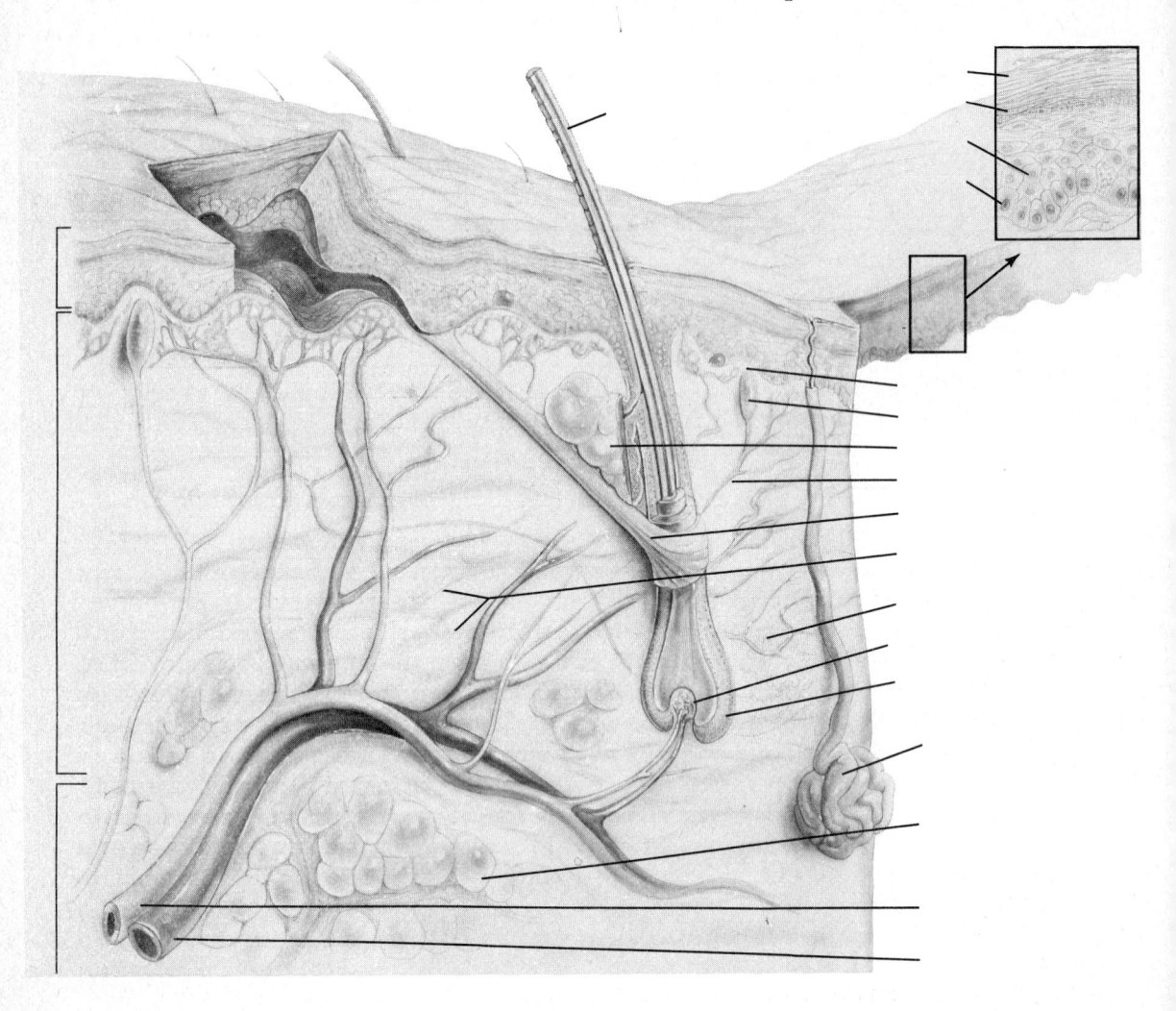

TEST YOURSELF

3. In thin skin the epidermal layer just beneath stratum corneum is stratum   a. basale   b. dermis   c. spinosum   (d.) granulosum   e. subcutaneum

4. Stratum spinosum is found in the (a.) epidermis   b. dermis   c. subcutaneous layer

52

LO 4: Trace a basal cell through the epidermal strata describing its differentiation and eventual fate.

As a basal cell moves outwards through the layers of the epidermis it becomes first prickle-shaped, then flattened as it differentiates to produce keratin. Eventually it loses its nucleus and other organelles and becomes filled with keratin. Then it dies, becoming scalelike and finally, sloughs off the skin surface.

Describe a basal cell as it passes through each layer
a. stratum basale _push daughter cells up to the other layers_
b. stratum spinosum _living cells are anterr to each other_
c. stratum granulosum _cells are away from the blood supply_
d. stratum corneum _flat dead cells filled with_ nAtSr

TEST YOURSELF

5. In stratum spinosum basal cells a. die b. multiply c. appear flat and granular d. differentiate and begin to produce keratin e. are scalelike

6. The youngest epidermal cells would be found in stratum a. basale b. spinosum c. lucidum d. corneum e. granulosum

LO 5: Describe the composition and functions of the dermis.

The dermis consists of connective tissue containing large amounts of collagen. It gives mechanical strength and substance to the skin, holds the blood vessels that nourish the epidermal cells and holds sensory receptors and epidermal structures such as hair follicles and glands.

1. Of what type of tissue is the dermis composed_____

2. What are dermal papillae_____

3. Where do the hair follicles and glands of the dermis originate_____

TEST YOURSELF

7. Which of the following is not normally found in the dermis   a. blood vessels   b. nerves   c. connective tissue   d. hair follicles   e. layer of dead cells

8. Which of the following is a function of the dermis   a. prevents water from entering the body   b. protects against ultraviolet rays   c. prevents dehydration   d. gives mechanical strength to the skin   e. acts as the first barrier to germs that may enter the body surface

> LO 6: Describe the composition and give the functions of the subcutaneous layer.

The subcutaneous layer consists of connective tissue, most notably fat. This tissue cushions underlying structures against mechanical injury, connects skin with tissues beneath, and stores energy in the form of fat.

Imagine that the subcutaneous layer suddenly disappeared from your body. What would be the consequences

_____
_____

TEST YOURSELF

9. The characteristic tissue of the subcutaneous layer is   a. adipose tissue   b. squamous epithelium   c. cuboidal epithelium   d. muscle   e. nervous tissue

10. Which of the following is NOT a function of the subcutaneous layer   a. protects against mechanical injury   b. connects skin with tissue beneath   c. stores energy in form of fat   d. waterproofs the skin

> LO 7: Describe the origin and structure of hair and tell how hair grows.

Hair follicles develop from epidermal cells that push down into the dermis. Like epidermis, a hair consists of cells that proliferate, differentiate, and then die as they move outward. The shaft of the hair consists of dead cells. Each follicle goes through a growing phase followed by a resting phase; then a new hair begins to grow pushing the old one out.

1. On a separate sheet of paper, draw and label a diagram of a hair.
2. Why can you cut hair without feeling pain_____
3. What are arrector pili muscles_____
_____

TEST YOURSELF

11. The part of the hair that we see is the   a. papilla   b. shaft   c. follicle   d. root   e. capillary

12. The epithelial cells that multiply giving rise to the cells of the hair are derived from the   a. subcutaneous layer   b. collagen of the dermis   c. stratum basale   d. stratum corneum   e. muscle tissue

> LO 8: Give the origin of sebaceous glands and describe the secretion of sebum; relate acne and comedos to malfunction.

Sebaceous glands are outgrowths of the same cells that form the hair follicles and are attached to hair follicles by ducts. They are holocrine glands which produce sebum from their own decomposing cells. Acne and comedos are skin lesions associated with malfunction of sebaceous glands.

1. What is sebum_____
2. Why is acne common during puberty_____

TEST YOURSELF

13. Sebaceous glands develop from   a. epidermal cells   b. collagen fibers in the dermis   c. adipose tissue of the subcutaneous layer   d. underlying muscle   e. epithelial cells within the dermis

14. A comedo is formed when a. the duct of the sebaceous gland and hair follicle is obstructed by accumulated sebum b. a duct ruptures allowing bacterial to spill into the dermis c. a hair follicle is destroyed d. acne goes untreated

> LO 9: Describe the homeostatic action of sweat glands in maintaining normal body temperature.

Sweat glands release a dilute salt water onto the skin surface which evaporates, cooling the body. When body temperature rises activity of the sweat glands increases thereby bringing the body temperature back towards normal.

1. How is sweat formed_____

2. Under what conditions does sweating increase _____

3. How would homeostasis be affected if suddenly all sweat glands were to stop functioning_____ _____

TEST YOURSELF

15. You are playing a fast game of tennis and a great deal of heat is generated by your muscular activity. To preserve homeostasis the activity of your sweat glands will a. decrease b. increase c. stay the same

> LO 10: Recount the locations and give the function of apocrine sweat glands.

Apocrine sweat glands are located in armpits and genital areas; they are responsible for body odor.

What causes the odor of the sweat from the apocrine sweat glands _____

16. Activity of the apocrine sweat glands is increased when one   a. is under emotional stress   b. sexually stimulated   c. asleep   d. answers (a) and (b) only are correct   e. answers (a),(b) and (c) are correct

> LO 11: Describe nails as outgrowths of epidermis

Nails are a modification of the horny epidermal cells and are composed of tough keratin.

TEST YOURSELF

17. Nails are composed of   a. melanin   b. keratin   c. sebum   d. collagen   e. adipose tissue

> LO 12: Describe the process of pigmentation and give the function of melanin in protection against ultraviolet radiation.

Melanocytes in the basal layer produce pigment granules and pass them on to other epidermal cells including those that form hair.   Melanin absorbs ultraviolet rays from the skin preventing damage to dermis and blood vessels.

1. What is melanin_____

2. Where are the melanocytes located_____
_____

TEST YOURSELF

18. Melanocytes   a. are found in the basal layer of the epidermis   b. inject pigment granules into other epidermal cells   c. are more active in dark skinned individuals   d. answers (a) and (b) only are correct   e. answers (a),(b) and (c) are correct

> LO 13: Explain how the skin responds homeostatically to excessive exposure to sunlight.

Exposure to the sun stimulates an increase in the amount and darkness of melanin providing a more efficient sunscreen to protect the body against ultraviolet radiation. We call this tanning.

1. In what way is tanning a protective response_____

2. What causes sunburn_____
_____

TEST YOURSELF

19. Exposure to the sun a. stimulates an increase in the amount of melanin produced b. causes the skin to become darker c. may cause a decrease in the number of melanocytes d. answers (a) and (b) are correct e. answers (a),(b) and (c) are correct

>LO 14: Describe how the skin may be used as an indicator of certain disease states.

Changes in the skin reflect internal disorders.

1. Briefly describe how the skin might reflect each of the following
   a. allergy_____
   b. anemia_____
   c. liver malfunction_____
   d. respiratory disorder_____

TEST YOURSELF

20. Pale skin may indicate a. anemia b. liver malfunction c. allergy d. respiratory disorder e. sunburn

>LO 15: Describe the process of inflammation and its characteristic symptoms, and explain why it is a homeostatic mechanism.

Inflammation is an attempt by the tissues to protect themselves from further injury or irritation. It is an attempt to destroy and dispose of harmful micro-organisms dilute harmful chemicals, clean up dead or injured cells, and wall off the injured area. Symptoms of inflammation are redness, swelling, heat, and pain.

1. How do histamines contribute to inflammation_____

2. What causes an inflamed region to feel warm and appear red_____
   _____

3. In what way is inflammation homeostatic_____
   _____

TEST YOURSELF

21. In inflammation the swelling is helpful because the excess fluid   a. dilutes harmful chemicals   b. reduces the amount of oxygen that can diffuse into the area   c. reduces the flow of nutrients into the area   d. answers (a),(b) and (c) are correct   e. answers (a) and (b) only are correct

22. Which of the following is NOT a characteristic of inflammation   a. the injured area is walled off   b. histamine is released   c. large numbers of white blood cells move into the area   d. the area feels cool to the touch   e. swelling and pain

LO 16: Describe the process of wound healing.

Cells multiply to fill in a wound and repair the damaged area. Healing may involve formation of a clot, scab, granulation tissue, and scar tissue.

1. What is a scab_____
2. What is granulation tissue_____
3. Describe the role of collagen in wound healing
   _____

4. What is a hypertrophic scar_____

5. What is a keloid_____
   _____

TEST YOURSELF

23. Formation of scar tissue is mainly a process of a. collagen synthesis and breakdown b. histamine release c. forming new sweat glands, hair follicles and sensory receptors d. multiplication of epithelial cells e. formation of adipose tissue

24. A large bulging scar that extends beyond the bounds of the wound is called a a. hypertrophic scar b. collagen scar c. granulation d. keloid e. scab

> LO 17: Distinguish between first, second, and third degree burns and tell how each is treated.

In first degree burns there is inflammation but the epidermis remains intact; in second degree burns epidermal cells die and inflammation and blistering occur; in third degree burns the damage may extend into the subcutaneous tissue exposing the victim to fluid loss and infection.

1. Explain how third degree burns dramatically point out the vital homeostatic functions of the skin. (What are the consequences upon homeostosis when one suffers extensive third degree burns)_____
_____
_____

TEST YOURSELF

25. In third degree burns a. treatment with cold packs is sufficient b. fluid loss is a serious problem c. sufficient numbers of epithelial cells remain so that regeneration quickly occurs d. no special treatment is required e. there is no need to apply the rule of nines .

> LO 18: Distinguish between benign and malignant neoplasms: List ways in which cancer cells differ from normal cells, and give the apparent causes and cellular basis for cancer.

The cells of a benign tumor tend to stay together and the tumor grows slowly. A malignant neoplasm is a cancer; its cells multiply rapidly and do not recognize boundaries of other cells.

1. List 3 apparent "causes" of cancer_____

2. What is metastasis_____
_____

TEST YOURSELF

26. Which of the following is <u>not</u> a characteristic of a malignant neoplasm a. metastasis b. invades neighboring tissues c. grows slowly d. triggered when DNA in a cell is mutated e. "caused" in some cases by excessive exposure to radiation

> *LO 19: Give the "causes" of skin cancer and identify three types discussed in the chapter.*

Skin cancer is caused by excessive exposure to the ultraviolet radiation from the sun, exposure to arsenic compounds, to x-rays, and to radioactive materials. Three types are basal cell carcinoma, squamous cell carcinoma, and malignant melanoma.

1. Skin cancer is most frequently "caused" by_____
_____

2. Describe basal cell carcinoma_____
_____

TEST YOURSELF

27. Cancer of the melanma   a. basal cell carcinoma b. malignant melanoma c. squamous cell carcinoma d. benign melanoma e. radiation carcinoma

ANSWERS TO TEST YOURSELF QUESTIONS

1.b   2.e   3.d   4.a   5.d   6.a   7.e   8.d   9.a
10.d 11.b 12.c 13.a 14.a 15.b 16.d 17.b
18.e 19.d 20.a 21.a 22.d 23.a 24.d 25.b
26.c 27.b

POST TEST

1. (LO 1) Which of the following are considered part of the integumentary system   a. hair   b. glands   c. skin   d. nails   e. all of the preceding answers are correct

2. (LO 2) Which of the following is NOT a function of the skin   a. helps maintain body temperature   b. prevents water loss   c. nutrition   d. prevents entrance of microorganisms   e. keeps us from getting waterlogged when swimming

3. (LO 3,4) New cells of the epidermis are formed in stratum   a. lucidum   b. granulosum   c. corneum   d. basale   e. spinosum

4. (LO 3) Stratum spinosum is located within the   a. dermis   b. epidermis   c. subcutaneous layer   d. basale   e. corneum

5. (LO 4) Intercellular bridges are characteristic of   a. basal cells   b. prickle cells   c. scales of stratum corneum   d. melanocytes   e. hair follicle cells

6. (LO 5) In the skin, nerves and blood vessels are found mainly in the   a. stratum granulosum   b. stratum corneum   c. stratum spinosum   d. dermis   e. stratum basale

7. (LO 6) Skin is attached to the underlying tissues by the   a. dermis   b. epidermis   c. stratum corneum   d. stratum basale   e. subcutaneous layer

8. (LO 7) Hair follicles a. are all formed before birth b. may be destroyed by electrolysis c. consist of dead cells and their products d. answers (a),(b) and (c) are correct e. answers (a) and (b) only are correct

9. (LO 8) Sebaceous glands are commonly known as a. pigment granules b. follicles c. oil glands d. sweat glands e. apocrine glands

10. (LO 7,8) Which of the following are found in the dermis but derived from the epidermis a. sebaceous glands b. hair follicles c. blood vessels d. answers (a) and (b) only are correct e. answers (a), (b) and (c) are correct

11. (LO 9) The skin structures that help regulate body temperature are the a. sweat glands b. sebaceous glands c. melanocytes d. hair follicles e. papules

12. (LO 10) The secretion of the apocrine sweat glands is odorous due to a. chemicals within the gland b. the type of secretion released c. the action of bacteria d. mixture with sebum e. the action of melanocytes

13. (LO 11) Which of the folowing is composed of tough keratin a. nails b. melanocytes c. hair follicles d. sweat gland ducts e. sebum

14. (LO 12) Ultraviolet rays from the sun are absorbed by a. melanin b. sebum c. hair papillae d. albinos e. scales in the basal layer

15. (LO 13) The amount of melanin in the skin increases when a. one is anemic b. one is an albino c. the skin is exposed to sunlight d. body temperature decreases e. hair enters the resting phase

16. (LO 14) When skin appears bluish in color which of the following may be indicated a. respiratory disorder b. anemia c. allergy d. melanin stimulation e. liver malfunction

17. (LO 15) When histamines are released a. blood vessels dilate b. blood vessels become more permeable c. large amounts of fluid may leave the blood and enter the inflamed area d. answers (a), (b) and (c) are correct e. answers (a) and (b) only are correct

18. (LO 16) During wound healing collagen is produced by a. epithelial cells b. melanocytes c. fibroblasts d. granulation tissue capillaries e. keloids

19. (LO 17) Susan has just suffered a burn. Her wound is inflamed but the epidermis remained intact. Her burn may be described as a. first degree b. second degree c. third degree d. hypertrophic

20. (LO 18) Jim has a neoplasm which has metastasized. He has a. cancer b. a malignant neoplasm c. a benign neoplasm d. basal cell carcinoma e. two of the preceding answers are correct

21. (LO19) A type of skin cancer in which cells of the lowest layer of epidermis lose their ability to form keratin and migrate into the dermis and subcutaneous tissues is known as a. malignant melanoma b. squamous cell carcinoma c. basal cell carcinoma e. benign neoplasm

ANSWERS TO POST TEST QUESTIONS

1.e   2.c   3.d   4.b   5.b   6.d   7.e   8.e   9.c
10.d   11.a   12.c   13.a   14.a   15.c   16.a   17.d
18.c   19.a   20.e   21.c

5 THE SKELETON

PRETEST

1. The main function of the skelton is to _transmit_ muscular force, but it also serves as a _calcium_ storehouse, for the _protection_ of internal organs and for the general _support_ of the body

2. A muscle is able only to _contract_ (give action). Therefore they are arranged in such a way as to _antagonize_ one another's actions

3. Most muscles may be said to have a less moveable _insertion_ and a more moveable _origin_. The _insertion_ is generally the more distal of the two

4. Muscles are often attached to bones by cable-like _tendons_

5. Describe each of these actions
   Flexion _____
   Extension _____
   Abduction _____
   Pronation _____
   Inversion _____

6. The ends of long bones are generally equipped with smooth plates of hyaline articular _cartilage_

7. Bones are covered with a thin connective tissue membrane called the _periosteum_, one of whose functions is to supply bone with _blood_
8. A long bone contains a central _marrow_ cavity lined with a delicate membrane, the _endosteum_
9. The three main parts of a growing long bone are the epiphysis, the _metaphysis_ and the _diaphysis_
10. The marrow cavity contains _cancellous_ bone
11. _Endochondral_ bone develops in the embryo from an original cartilage
12. Bone-forming cells are known as _osteoblasts_
13. Bony spicules in developing membrane bone spread from the _center_ to the _border_
14. The cartilage of endochondral bone becomes invaded by blood vessels after it has become _calcified_
15. The osteoblast's basic task is to secrete _collagen_ which chemically is a _protein_
16. Osteocytes live in tiny _lacunae_ in the bone
17. Osteoclasts are multinucleated, giant cells that _destroy_ bone
18. Long bones are able to grow in length thanks to their _metaphyses_
19. The earliest type of bone laid down in an endochondral bone is _woven bone_
20. Woven bone is usually replaced by _lamellar_ bone
21. Describe an osteon _consists of a gro of lamellar bone deposit_
22. A _blood clot_ forms around a broken bone, which is then invaded by osteoblasts which produce a splintlike _callus_
23. What disease involves an infection of the bone _infection of bone_
24. What disease replaces lamellar bone with woven bone _Paget's disease_
25. Hyaline cartilage occurs at the ends of _long bones_
26. Cartilage grows by a process called _instration_ growth as well as by _____ growth resulting from the action of the _____ membrane
27. The membrane which covers some tendons is called the _peritendium_
28. Tendons attach to bones by means of _Sharpey fibers_

29. List the varieties of immoveable joints
    (a) _synedrosis_
    (b) _synthose_
    (c) _syndesmoses_
    (d) _symples_

30. A diarthrodial joint is surrounded by a connective tissue _capsule_ which encloses the _synovial_ cavity, and which secretes the _synovial_ fluid

31. In the space provided diagram a hinge joint

32. _Rheumatoid Arthritis_ is a generalized disease which affects the joints, probably produced by a disorder of the body's _immune_ mechanism

33. In gout, deposits of _uric_ _acid_ frequently form in joints

ANSWERS TO PRETEST QUESTIONS

1. transmit- calcium (and/or phosphate)-protection- support (LO 1) 2. contract-antagonize (LO 2) 3. origin-insertion-insertion, in that order (LO 2) 4. tendons (LO 5) 5. Too lengthy for inclusion here. See textbook. (LO 2) 6. cartilage (LO 3) 7. periosteum- blood (LO 3) 8. marrow- (medullary) endosteum (LO 3) 9. metaphysis-(epiphyseal plate)- diaphysis, any order (LO 3) 10. cancellous (LO 4) 11. endochondral (LO 4) 12. osteoblasts (LO 5) 13. center-border (or edges), in that order (LO 5) 14. calcified (LO 5) 15. collagen-protein (LO 6) 16. lacunnae (LO 6) 17. destroy (LO 6) 18. metaphyses (LO 7) 19. woven bone (LO 7) 20. lamellar (LO 7 & 8) 21. Consult textbook. Too lengthy to include here. Your answer should however include the concept that an osteon consists of layers of lamellar bone deposited around a central blood vessel (LO 8) 22. blood clot-callus (LO 9) 23. an infection of a bone (LO 10) 24. Paget's disease (LO 10) 25. long bones (LO 11, 12)

26. interstitial-appositional-perichondral, in that order (LO 11) 27. peritendineum (LO 11) 28. Sharpey's fibers (or penetrating fibers)(LO 11) 29. syndesmoses, synostoses, synchondroses, symphyses; any order (LO 12) 30. capsule-synovial (or joint)-synovial (LO 12) 31. Consult Fig. 5-25 in the textbook (LO 12) 32. Rheumatoid arthritis-immune (LO 13)   33. uric acid (LO 13)

LO 1: Summarize the functions of the skeleton

The skeleton serves for the transmission of muscular forces, for calcium and phosphate storage, for support and protection, and for the production of blood cells.

The major functions of the human skeleton are (1)_____
(2)_____ (3)_____ (4)_____

TEST YOURSELF

1. The skeleton functions a. as a calcium store b. to transmit forces c. for support of body soft tissues d. for the protection of internal organs e. in all of the preceding ways

2. Summarize the functions of the muscular system in relation to the skeleton.

Muscles may contract, thus serving for locomotion and for internal functions such as the movement of the gastrointestinal tract. They are arranged in opposing pairs or groups, called antagonists, with the skeletal system generally serving to transmit forces between and among them.

1. The basic muscular action is_____
2. How can a muscle act as part of the skeleton_____
_____

3. A muscle that produces an action is called an
_____:
one that produces the opposite action is its
_____
4. What is (a) opposition_____
   (b) adduction_____
   (c) abduction_____

68

5. Why is it advantageous for extrinsic muscles to move the thumb by means of tendons_____
_____
_____

TEST YOURSELF

2. Muscles may actively a. contract b. expand c. (a) and (b) d. undertake a range of actions impossible to describe briefly

3. In many cases muscular forces are expressed at a distance by means of a. the circulatory system b. tendons c. articular cartilages d. ligaments

4. An agonist a. is a sense organ that responds to pain b. produces the opposite effect c. is responsible for some action d. is a special muscle that opposes the thumb to the other fingers e. moves a body part away from the main axis (or away from the main axis of a limb)

> LO 2: Describe the general characteristics of a typical bone.

A typical bone is surrounded by a periosteal membrane, possesses an outer crust of compact bone and an inner filling of cancellous bone and marrow. In the case of long bones, there is generally a synovial joint at each end, and the bone itself consists of diaphyseal, metaphyseal and epiphyseal portions. The metaphysis permits lengthwise growth of long bones; the periosteum, growth in diameter. Membrane bones grow generally at their edges.

1. At their ends, long bones possess surfaces of _____ cartilage
2. The remainder of the bone is covered with a membrane, the _____
3. The periosteum, among other functions, supplies the bone with_____
4. The periosteum is fastened to the bone by means of _____   _____

5. The marrow cavity is loosely packed with_____ bone and lined with the delicate_____ membrane
6. The main shaft of a long bone is the_____; the very ends are the_____and the growing, cartilagenous discs between these are the
_____

TEST YOURSELF

5. The periosteum a. is composed of epithelium b. is penetrated by blood vessels on their way to the marrow cavity c. forms articular surfaces on the bone ends d. lines the marrow cavity

6. Growth in bone length is a function of the a. periosteum b. endosteum c. epiphysis d. diaphysis e. metaphysis

> LO 4: List and describe the main classifications of bone, i.e., by shape, texture, tissue organization, and embryonic origin.

Bone is classified variously by shape (mainly long, flat tabular and irregular) by tissue organization (woven vs. lamellar) and by development (membranous vs. endochondral).

1. Refer to fig._____ and classify each of the following bones as "long" or "tabular" (a) femur_____
   (b) frontal_____ (c) innominate_____
   (d) phalanges_____
2. Most bones possess an outer shell of _____ bone
3. An endochondral bone begins as a model composed of _____cartilage replacement_____; while a membrane bone develops by laying down bony spicules in a membrane composed of ___connective tissues___

TEST YOURSELF

7. The bones of the limbs are for the most part _____ bones a. irregular b. tabular c. long d. membrane e. impossible to say

8. Bone that begins as a model in cartilage is known as _____ bone   a. endochondral   b. perichondral   c. periosteal   d. membranous   e. cancellous

> LO 5: *Describe the development of membranous bone and endochondral bone.*

Membranous bone develops in plaques of connective tissue, whereas endochondral bone is the result of cartilage replacement.

1. Membrane bones begin in flat membranes composed of
2. Bony _____ appear in the centers of these membranes, laid down by bone-forming cells called
3. _____ bone begins as a model made of
4. What part of a long bone is the LAST to calcify
5. As calcified cartilage is invaded by blood vessels, it begins to be destroyed, and in pace with this destruction, bone is laid down by _____
6. Describe the fashion in which membrane bones grow
   _____
   _____
   _____

TEST YOURSELF

9. Which of the following do membrane and endochondral bone have in COMMON   a. spicules   b. cartilage   c. bone formation by osteoblast action   d. epiphyses   e. fontanelles

> LO 6: *Give the functions of osteoblasts, osteoclasts and osteocytes.*

Osteoblasts lay down collagen fibers which serve as nuclei for mineralization. When they become trapped in forming bone, they are transformed into osteocytes, which occupy lacunae in the mature bone. Osteoclasts are multinucleated cells which break down existing bone, mainly during skeletal remodeling. NOTE: Very recent research indicates that in the reabsorption of dead and diseased bone, and possibly also in normal remodelling, monocytes and macrophages also destroy bone by means of some secreted substance, possibly a collagenase enzyme.

1. The basic mission of the osteoblast is to secrete _____, originally produced, however, as the nonfibrous protein_____
2. Collagen lends_____to bone, and serves as a nucleus for the deposit of the mineral_____
3. Osteocytes originate as trapped_____ which become isolated in little spaces called_____
4. Cells which break down bone include the_____, the_____and the_____. The _____are giant, multinucleated cells

TEST YOURSELF

10. Bone tissue formation begins when osteoblasts deposit   a. apatite   b. calcium   c. collagen   d. silica   e. lysosomes

11. Osteocytes originate from   a. osteoclasts   b. trapped osteoblasts   c. white blood cells   d. chondrocytes   e. megakaryocytes

12. Collagen serves as   a. a component of the skin   b. as a structural part of bone tissue   c. nuclei for apatite crystallization   d. a component of tendons   e. all of the preceding

LO 7: Describe bone growth and remodeling

Bone initially forms as woven bone, which has no grain. Woven bone is rapidly replaced by lamellar bone, whose collagenous and crystalline structure is highly oriented. Both cancellous (spongy) and compact types of bone are generally composed of lamellar bone. Osteoblasts, in conjunction with the periosteum, lay down new bone, causing the involved bone to grow, at least in diameter. Meanwhile, osteoclasts remove bone from the interior and reorient the grain of existing bone through the formation of osteons (haversian systems).

1. In order to save weight and change bone structure in accordance with the changing needs of the organism, bone is extensively _____
2. Osteoclasts are giant cells with multiple _____
3. _____ bone, which has little definate structure, is laid down before lamellar bone
4. It is believed that woven bone reacts electrically to stress, guiding the orientation of the _____ bone that replaces it
5. _____ disease converts lamellar to woven bone

TEST YOURSELF

13. Which of these forms LAST   a. lamellar bone   b. woven bone   c. cartilage model   d. connective tissue plaque

14. Osteoclasts   a. secrete collagen   b. secrete tropocollagen   c. destroy collagen   d. produce cartilage   e. secrete alkaline phosphatase enzyme

> LO 8: Describe the structure of the osteon (haversian system) and relate this to its function.

The osteon is composed of several layers of bone containing spirally wound collagen fibers at mutual angles to one another. Osteocytes are found in lacunae located between the layers and the whole unit forms around a blood vessel. By interdigitating with one another, osteons lend a grain to bone which enables it to withstand deforming forces better than woven bone would be able to.

1. Label the accompanying diagram

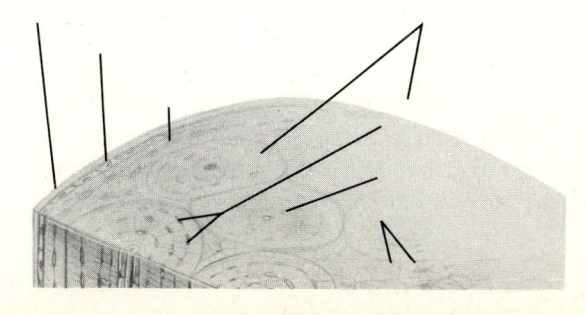

2. What function are osteons believed to serve_____
   _____
   _____
   _____

TEST YOURSELF

15. An osteon a. surrounds a tiny blood vessel b. is made up of concentric layers of bone c. replaces woven bone d. replaces lamellar bone e. all of the preceding

> LO 9: Describe the healing of bone fractures

In general, when a bone is fractured it becomes surrounded by a clot of blood. This clot is soon invaded by osteoblasts from the periosteum, which proceed to form a callus of woven bone about the fracture. Remodeling processes then unite the broken ends.

1. The first step in the repair of a fractured bone is the formation of a _____ (known otherwise as a hematoma)

2. Next, the adjacent periosteum contributes_____ which invade the hematoma and convert it to a splintlike_____of woven bone

3. The callus may eventually be reabsorbed and the original shape of the bone restored by skeletal
   _____

TEST YOURSELF

16. The cells surrounding a break in a bone which begin to form the callus are derived from a. the metaphysis b. a blood clot c. the periosteum d. osteocytes of the broken edges e. (b) and (c)

> LO 10: List and describe four important diseases of bone.

(For a summary, consult table 5-1 in the textbook).

(1) _____
(2) _____
(3) _____
(4) _____

TEST YOURSELF

17. Infection of the periosteum and marrow cavity
18. Bone becomes spongy and decalcified
19. Softening and bowing of long bones
20. Bone cancer of young people

a. osteosarcoma
b. Paget's disease
c. osteomyelitis
d. rickets
e. osteoporosis

>   LO 11: Describe the structure and summarize
>   the function of tendons and cartilage
>   (consulting chapters 1 and 2 if necessary)

Ligaments are cables or sheets of inelastic connective tissue which may bind bones together at joints. Tendons serve to transmit forces from muscle to bone. A tendon may be enveloped in a sheath, the peritendineum, and this in turn may be surrounded by a bursa. The bursa is an extension of the synovial cavity of an adjacent joint, and contains synovial fluid.

Cartilage comprises hyaline, elastic fibrous, and inelastic fibrous varieties. Usually surrounded by a perichondrium, cartilage contains more glycoprotein ground substance than collagen.

1. Cartilage is surrounded by a connective tissue membrane called the_____and tendons, similarly, are often surrounded by a_____
2. The lacunar cells of cartilage are called_____
3. Cartilage is capable of both_____growth and_____growth, unlike bone which only employs_____growth
4. Ligaments are derived from the walls of the joint _____and attach_____to one another

5. Tendons attach bones to _____
6. The_____serving as tendon envelopes, are often connected with the joint cavities
7. The tendon's end is attached to the bone by _____fibers, which as you will recall, also attach the_____to the bone surface.

TEST YOURSELF

21. tendon covering
22. tendon sheath
23. unites bone
24. contains synovial fluid

a. bursa
b. ligaments
c. peritendineum
d. periosteum
e. perichondrium

LO 12: List the main types of joints, describe their structure, and give their function. Classify examples.

LO 13: Give the principal joint disorders

(See figure 5-25 in the text for a diagrammatic summary of the types of joints. Consult various illustrations in chapters 6 and 7 for examples.)
    Some of the main joint disorders are dislocations, in which the synovial cavity may be enlarged or ruptured by trauma and in which articulation is lost; infectious arthritis; osteoarthritis; and rheumatoid arthritis. The arthritic disorders are inflammations of the joint cavity.

1. Any connection between bones is a_____
2. Of the immoveable joints, those united by fibrous connective tissue are termed_____; those connected by hyaline cartilage are called _____; and those in which bones are directly fused at their edges are called _____. Symphyses, the most flexible of them are composed of_____ _____
3. In diarthrodial joints, the associated bones may be _____
4. The perichondrium is missing from_____ cartilages
5. Articular cartilages are lubricated by_____ fluid

6. In each case below, label the joint type

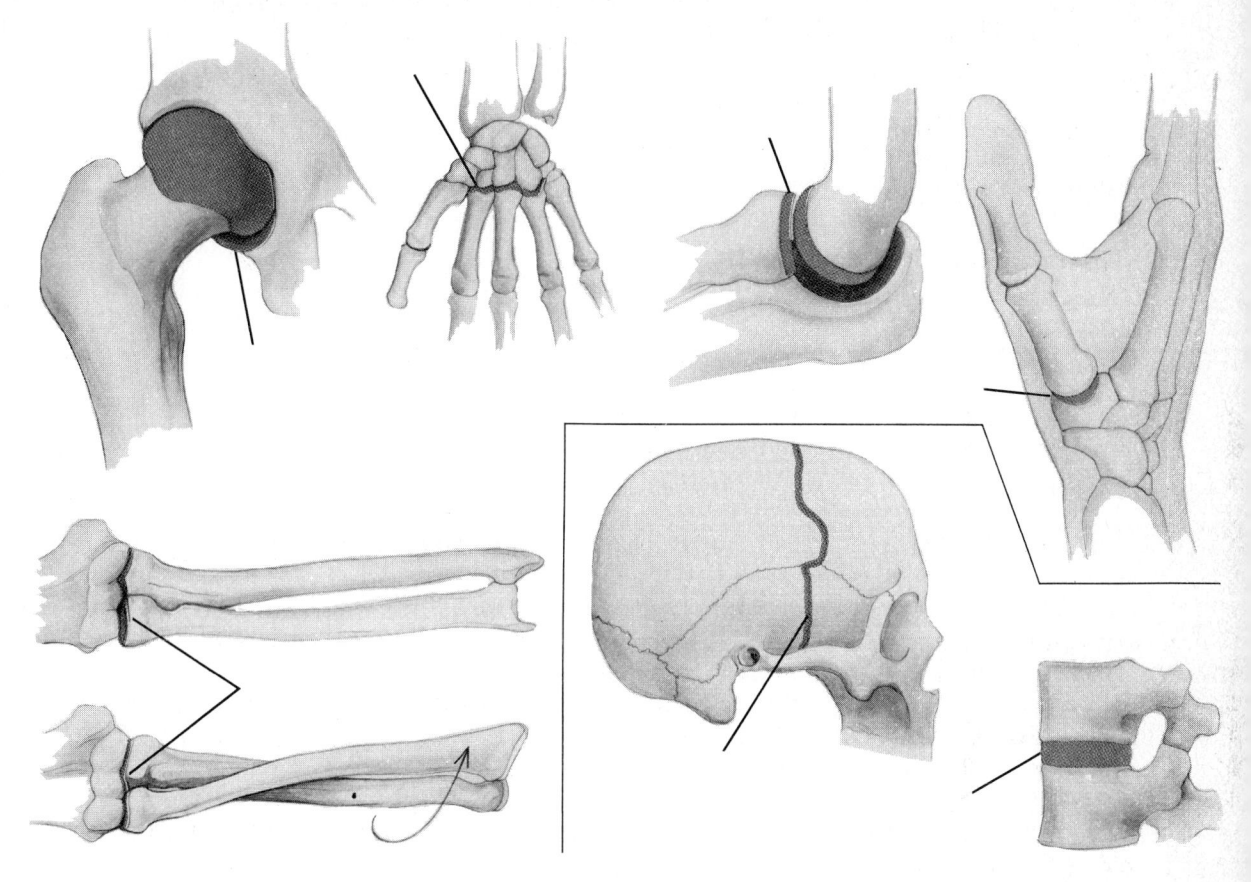

7. Bacteria are responsible for _____ arthritis
8. _____ is the joint aspect of a generalized inflammatory disease whose cause is unknown
9. In gout, deposits of _____ acid called _____ may interfere with joint action

TEST YOURSELF

25. interphalangeal finger joints
26. synostosis
27. hip joint
28. symphysis
29. synchondrosis

a. hinge joint
b. ball-and-socket joint
c. joint formed by bone fusion
d. most moveable of the non-synovial joints
e. none of these

30. An arthritic condition found in old age, evidently caused by a wearing away of the articular cartilages is called   a. rheumatoid arthritis   b. osteoid arthritis   c. gouty arthritis   d. infectious arthritis   e. Paget's disease

ANSWERS TO TEST YOURSELF QUESTIONS

1.e	2.a	3.b	4.c	5.b	6.e	7.c	8.a	9.c
10.c	11.b	12.e	13.a	14.c	15.e	16.c	17.c	
18.e	19.d	20.a	21.e	22.a	23.b	24.a	25.a	
26.c	27.b	28.d	29.e	30.b				

POST TEST

1. (LO 1) Which of the following is NOT a skeletal function   a. support   b. protection   c. potassium storehouse   d. transmission of forces   e. all of the preceding are, in fact, skeletal functions

2. (LO 2) The antagonistic arrangement of muscles is a consequence of the fact that   a. muscles push bones   b. a muscle can only contract   c. there isn't enough room in appendages for the muscles that control them   d. complex movements require the participation of many muscles   e. none of the preceding reasons

3. (LO 2) spreading the fingers
4. (LO 2) role of the thumb in the power grip of the hand
5. (LO 2) touching the right shoulder with the right hand
6. (LO 2) straightening the leg

a. opposition
b. extension
c. abduction
d. supination
e. flexion

7. (LO 3) The ends of a long bone are made of   a. cancellous bone   b. cartilage   c. periosteum   d. fibrocartilage   e. dense bone

8. (LO 4) The spongy bone within the marrow cavity of the mature radius should be classified as   a. membrane   b. periosteum   c. fibrocartilage   d. cancellous bone   e. woven bone

9. (LO 5) Which portion of an endochondral bone is the last to ossify   a. diaphysis   b. metaphysis   c. epiphysis   d. edge   e. center

10. (LO 6) When an osteoblast becomes isolated in a lacuna, it is transformed into a(n)   a. osteocyte   b. osteoclast   c. Sharpey's fiber   d. fibroblast   e. macrophage

11. (LO 6) An osteoblast produces collagen fibers from   a. enzymes in its lysosomes   b. tropocollagen   c. glycogen   d. apatite   e. mucus

12. (LO 7) Lamellar bone   a. forms directly from cartilage   b. has no definite grain   c. replaces woven bone   d. is the opposite of cancellous bone   e. occurs only in tabular bones such as the flat bones of the skull and pelvis

13. (LO 8) An osteon consists of concentric layers of lamellar bone arranged around a   a. nerve   b. lacuna   c. collagen fiber   d. blood vessel   e. tendon

14. (LO 8) An osteon   a. is spindle-shaped   b. may be replaced by other osteons   c. consists of layers whose grain lies at mutual angles   d. arises from the endosteum   e. all of the preceding

15. (LO 9) When a bone fracture heals, the initial blood clot is first replaced with   a. lamellar bone   b. hyaline cartilage   c. connective tissue   d. woven bone   e. fibrous cartilage

16. (LO 10) Osteomyelitis is   a. bone infection often occuring in young people   b. a bone cancer   c. an inflammatory joint disease   d. wear and tear of the articular cartilages often found in older people   e. the pathologic replacement of lamellar bone with woven bone

17. (LO 11) Cartilage differs from bone in that it a. is capable of interstitial growth b. usually has no blood vessels c. has no channels between the lacunae d. has relatively little collagen e. all of the preceding, or all of the preceding except (d)

18. (LO 11) (IN THIS QUESTION INDICATE THE INCORRECT CHOICE!) Tendons a. are attached to bone by Sharpey's fibers b. transmit the pull of muscles c. consist largely of collagen fibers d. may possess sheaths called bursae e. contain cells similar to chondrocytes in interconnecting lacunae

19. (LO 12) The bones of the skull fuse in middle age, producing joints called a. synchondroses b. synostoses c. symphyses d. gomphoses e. diarthroses

20. (LO 12) Synovial fluid is secreted by a. the articular cartilages b. the bursae c. the capsular villi d. the ligaments e. endosteal blood vessels

21. (LO 13) Uric acid may be deposited in the joint cavity in connection with the disease a. osteomyelitis b. osteosarcoma c. osteoarthritis d. Paget's disease e. gout

ANSWERS TO POST TEST QUESTIONS

1.c	2.b*	3.c	4.a	5.e	6.b	7.b	8.d	9.b
10.a	11.b	12.c	13.d	14.e	15.d	16.a	17.e
18.e	19.b	20.c	21.e

---

* a case could be made also for (d)

6 THE MUSCLES

PRETEST

1. As a rule, a muscle has a more moveable attachment, its_____ ____, and a less moveable_____
2. The entire muscle is surrounded by a connective tissue capsule called the_____. Internally, this tissue is connected to partitions called_____. The muscle cells themselves are individually wrapped in the _____
3. The contractile units of the muscle cell are called the_____fibrils
4. Striated muscle cells each possess_____nuclei
5. The several muscle cells controlled by a single nerve fiber collectively known as a _____ _____
6. The motor end plate is the junction between a _____and a muscle cell
7. However, there is no direct connection between the two, so that a transmitter substance,_____, must diffuse across the _____ _____ to initiate muscle contraction
8. Depolarization spreads within the muscle cell via the _____system
9. In response, the vesicles of the sarcoplasmic reticulum release_____which triggers muscle contraction. Afterwards the sarcoplasmic reticulum_____ this substance
10. Repeated closely successive stimuli produce a sustained muscular contraction called_____
11. The "thick" filaments of a muscle cell are composed of the protein_____whereas the thin filaments are made of_____

12. A unit of thick and thin filaments comprises a _____
13. Sarcomeres are joined by interwoven filaments comprising the _____-line
14. The cross-bridges between the thick and thin filaments are actually the ends of _____ molecules
15. The immediate source of energy for muscular contraction is the compound ATP. More long-term energy storage is the task of the related substance _____
16. Since skeletal muscle is often called upon to rely on the inefficient anaerobic type of metabolism as an energy source, it contains fairly large amounts of the stored cellular fuel _____
17. The anaerobic metabolism of glycogen by muscle builds up an oxygen_____ in the exerciser
18. _____ is a red pigment within the muscle cell that transfers oxygen internally
19. In the space below, describe the role of filaments and cross-bridges in producing the force and movement of muscular contraction. Be sure to include reference to troponin-tropomyosin, sites of attachment, and conformational change

20. Rigor-mortis refers to rigid muscular contraction occuring after_____
21. Myasthenia gravis is primarily a disease of the _____
22. Smooth muscle is generally voluntary/involuntary, lacks visible_____ and is also known as _____ muscle

82

ANSWERS TO PRETEST QUESTIONS

1. insertion-origin (LO 1) 2. epimysium-perimysium-endomysium (LO 1) 3. myo- (LO 1) 4. many (LO 1) 5. motor unit (LO 2) 6. a nerve fiber (or nerve cell) (LO 2) 7. acetylcholine-synaptic cleft (LO 2) 8. T (LO 2) 9. calcium-reabsorbs (LO 2) 10. tetany (LO 2) 11. myosin-actin (LO 3) 12. sarcomere (LO 3) 13. Z (LO 3) 14. myosin (LO 3) 15. ATP-CP (LO 3) 16. glycogen (LO 3) 17. debt (LO 3) 18. myoglobin (LO 3) 19. too lengthy for inclusion here. See textbook. (LO 3) 20. death (LO 3) 21. motor and plate (LO 4) 22. involuntary-striations-visceral (LO 5)

> LO 1: Describe the gross and microscopic structure of striated muscle.

A gross muscle is infiltrated with an elaborate connective tissue harness, which may be subdivided into the external epimysium, the perimysium system of internal partitions, and the endomysium surrounding the individual muscle cells. The muscle cells themselves are long, multinucleated, and subdivided into sarcomeres. Each sarcomere is a contractile unit containing thick (myosin) filaments and thin (actin) filaments which by pulling themselves past one another via cross bridges, produce contraction. Though not parts of the muscle itself, the origin serves as the less moveable and usually more proximal point of muscle attachment, and the insertion as the more moveable and usually distal point of attachment.

1. Where is the biceps muscle located_____
2. What is an origin_____
   an insertion_____
3. The perimysium is the _____ covering of a gross muscle, and the _____ is the connective issue wrapping of the individual muscle cells
4. The _____ divides the tissue of a muscle into bundles called _____
5. The multinucleated muscle cell arrives by the fusion of _____ cells

83

6. The _____ are physically responsible for the contraction of the muscle cell. They are organized into intracellular cylindrical groups called

7. The _____ in the muscle cell equivalent of the plasma membrane, and its equivalent of the ER is called the _____

TEST YOURSELF

1. wraps individual muscle cells
2. bundles of muscle cells
3. physical basis of muscle contraction
4. skeletal muscle markings
5. muscle cell equivalent of the ER

a. sarcoplasmic reticulum
b. filaments
c. endomysium
d. striations
e. fasciculi

LO 2: Describe the mechanisms of synaptic transmission, depolarization and calcium release as they relate to the control of muscle contraction.

When a nerve impulse reaches an end plate, acetylcholine is released into the synaptic cleft, where it is quickly destroyed by the enzyme cholinesterase. Before it is actually destroyed, acetylcholine depolarizes the muscle cell membrane. This area of depolarization now travels down the muscle cell and, as it passes the entrances to the T system, enters the muscle cell, where it causes the vesicles of the sarcoplasmic reticulum to emit calcium. The calcium stimulates the contraction of the muscle, by exposing the active sites of the thin filaments, and when reabsorbed, permits muscle relaxation.

1. A _____ is a group of muscle cells controlled by a single _____
2. The motor end plate is the _____ between nerve and _____ cell
3. The _____ substance released from the nerve cell ending is _____. This diffuses across the _____
4. In response, the muscle cell membrane displays a self-propagating electrical disturbance called the _____

5. Leftover acetylcholine is _____ by the enzyme _____
6. Depolarization is conveyed inside the muscle cell by the ____-system. In response, the _____ of the sarcoplasmic reticulum release _____
7. The filaments respond to this by _____ Afterwards, the calcium is _____ by an _____ mechanism into the vesicles of the _____ reticulum
8. A single, sharp muscular contraction is called a _____, but a smooth sustained contaction is called a _____ contraction
9. The strength of a muscular contraction depends mainly on the _____ of motor units employed
10. _____ pesticides interfere with the normal _____ of acetylcholine. On the other hand, botulinus toxin interferes with the _____ of acetylcholine

TEST YOURSELF

The following are steps in the contraction of a muscle cell   a. calcium release   b. acetylcholine release   c. depolarization of sarcolemma   d. depolarization of T-system

6. Of those listed, which is the FIRST step
7. With which step does Botulinus toxin interfere
8. Which step immediately precedes physical contraction
9. Which step precedes (a)

10. Acetylcholine is disposed of by   a. reabsorption by the nerve ending   b. reabsorption by the vesicles of the sarcoplasmic reticulum   c. destruction by cholinesterase   d. depolarization   e. none of the preceding

> LO 3: Describe the main mechanical and chemical aspects of striated muscle contraction.

The active interdigitation (by which we mean dovetailing) of the thin actin and thick myosin filaments of a sarcomere is powered by ATP associated with the myosin cross-bridges between the two kinds of filaments. This

ATP obtains its immediate energy from a pool of creatine phosphate, which in turn is synthesized using ATP; thus the creatine phosphate serves as an energy bank.

1. The two types of muscle filaments are the_____ and the_____
2. The _____ filaments are composed of the protein actin and the_____ filaments are composed of _____
3. The cross-bridges are the ends of _____ molecules
4. During contraction, each cross bridge attaches to an_____ _____ on the actin molecules. Successive attachments produce the actual _____
5. _____ provides the immediate power for cross bridge attachment and release
6. The "storage battery" or "bank" of muscular energy is the compound _____
7. The energy produced in muscle cells originates from the metabolism of _____ or from _____
8. When does a muscle cell respire anaerobically _____ _____
9. The head of the myosin cross-bridge moves by means of a _____ change in the protein of which it is composed
10. ATP is also required to detach the _____ _____ from the active site

TEST YOURSELF

11. Muscle contraction occurs when   a. myosin is converted into actin   b. glucose is converted into muscle glycogen   c. actin and myosin slide between one another   d. smooth muscle is transformed into skeletal muscle   e. all the preceding are correct

12. The energy for muscle contraction comes directly from   a. actin   b. myosin   c. shivering   d. movement of the myofilaments   e. ATP

13. Calcium acts upon the filaments by   a. uncovering the active sites   b. lubricating the cross bridges   c. breaking down ATP   d. stimulating aerobic metabolism   e. paying back the "oxygen debt"

14. Rigor mortis results from a. ATP deficiency b. postmortem changes c. irreversible union of cross bridges with actin d. all of the preceding

> LO 4: List the main varieties of muscle disease.

Muscles may become physically injured or infected, or they may suffer from a fatty degeneration of genetic origin (muscular dystrophy) or from a neuromuscular defect of unknown origin (myasthenia gravis). Muscles atrophy with age, and enlarge (because of the enlargement of individual cells) with exercise.

1. If the length, but not the tension of a muscle is changed, the type of contraction this represents is called_____contraction
2. Give an example of isometric contraction_____
_____
3. When a muscle enlarges as a result of exercise, it is due to_____of the muscle cells
4. In_____ _____ the muscles progressively atrophy
5. MG is basically a disease of the_____ _____
6. In_____ _____ the muscle degenerates from lack of nervous stimulation
7.

TEST YOURSELF

15. MG
16. Isometric
17. Trichinella spirillum
18. Muscle cell enlargement

a. parasitic worms
b. contraction without change in length
c. result of exercise
d. end plate disorder
e. irreversible contraction occuring after death

> LO 5: Describe the structure and action of smooth muscle, comparing it with striated muscle.

Smooth muscle consists of individually nucleated cells very closely attached to one another. Smooth muscle, though slower acting than skeletal, is capable of far greater contraction.

1. A smooth muscle cell possesses _____ nucleus(nuclei)
2. It is believed that impulses triggering contraction may pass directly between _____ _____ cells
3. In the space below, describe the innervation of smooth muscle cells _____
   _____
4. Smooth muscle _____ in response to stretching
5. The exact contractile mechanism of smooth muscle is _____

TEST YOURSELF

19. Smooth muscle contraction is known to contain a. actin b. myosin c. (a) and (b) d. glycogen e. none of the preceding

20. Smooth muscle will contract in response to a. stretching b. nerve action c. various transmitter substances and hormones d. impulses passed from other smooth muscle cells e. all of the preceding

ANSWERS TO TEST YOURSELF QUESTIONS

1.c   2.e   3.b   4.d   5.a   6.b   7.b   8.a   9.d
10.c 11.c 12.e 13.a 14.d 15.d 16.b 17.a
18.c 19.e 20.e

POST TEST

1. (LO 1) Myosin and actin a. are two proteins found in muscle b. make up the coarse and fine filaments of the myofibrils c. are energy molecules found in muscle d. answers (a),(b) and (c) are correct e. only answers (a) and (b) are correct

2. (LO 1) Cross bridges a. are made of actin b. are made of ATP c. are made of myosin d. occur only in

visceral muscle   e. attach muscle cells to one another to form fasciculi

3. (LO 1) Connective tissue is attached to the muscle cell by means of   a. the perimysium   b. the Z-line   c. dimples in the sarcolemma associated with the contractile filaments   d. the T-system   e. the sarcoplasmic reticulum

4. (LO 1) The striations of voluntary muscle result from   a. connective tissue sheaths of the muscle cell   b. a complex interweaving of microtubules   c. overlapping and interconnected thick and thin filaments   d. the cross-bridges   e. TWO of the preceding

5. (LO 2) A synapse is found within   a. the motor end plate   b. the T-system   c. the myofibril   d. the perimysium   e. the epimysium

6. (LO 2) _____ the membranes of the T-system leads to muscle cell contraction   a. calcium release from   b. depolarization of   c. action of acetylcholine upon   d. stretching

7. (LO 3) Calcium produces   a. depolarization   b. potassium release from the vesicles of the SR   c. contraction   d. the uncovering of the active sites of the actin   e. TWO of the preceding

8. (LO 3) ATP is required   a. as a source of muscular energy   b. for the release of cross-bridges from actin   c. for conformational motion of the cross-bridges   d. for the "charging" of CP with energy   e. for all of the preceding

9. (LO 4) Anticholinesterase drugs and poisons   a. tend to promote muscular contraction   b. destroy or inactivate cholinesterase   c. prevent acetylcholine release   d. destroy acetylcholine receptors of the sarcolemma   e. (a) and (b)

10. (LO 4) _____ is an inherited disorder producing atrophy and fatty degeneration of the muscles a. muscular dystrophy b. myasthenia gravis c. denervation atrophy d. isotonic contraction e. trichinosis

11. (LO 5) Visceral muscle a. is multinucleated b. is automatic in function c. possesses prominent striations d. is individually innervated e. does not respond to transmitter substances

ANSWERS TO POST TEST QUESTIONS

1.e   2.c   3.c   4.c   5.a   6.b   7.e   8.e   9.e
10.a   11.b

7 THE MUSCULOSKELETAL SYSTEM

NOTE: Different instructors vary greatly in the degree and kind of emphasis accorded to the musculoskeletal system. Because of this, it did not seem practical to provide a thorough or detailed list of learning objectives in the textbook for this chapter. However, it is not practical to use a study guide such as this without learning objectives, so we have attempted to develop a set of intermediate difficulty for use here. The use of all possible learning objectives would, however, have lengthened this unit beyond reason so that it has been necessary to omit most of the more detailed ones even from this study guide. Such an approach requires modification of our usual format.

You must consult with your instructor to discover just what his or her learning objectives for chapter 7 are. They may be quite different from those employed here.

PRETEST*

1. The basic action of all muscles is _____
2. Most muscles are shaped in such a way that their fibers_____upon some tendon or insertion
3. _____ tend to be broader/narrower than insertions
4. What is a synergist_____
5. What is a condyle_____
_____

*Some of the more detailed learning objectives have been omitted from this pretest.

6. What is a meatus_____
   _____
   (be sure that you know how to distinguish it from a foramen)
7. a. Why is it hard to walk with a sore big toe
   _____
   b. with a sore little toe_____

8. The _____ of the foot acts as a shock absorber
9. The_____ membrane (or ligament) stretches between the tibia and the_____
10. In a _____ class lever the fulcrum is placed between a long force arm and a short_____ arm
11. Most limb bones, joints and muscles are so arranged as to comprise_____ class levers
12. In a third class lever,_____ is sacrificed to gain greater _____ _____
13. Most thigh muscles insert on the _____
14. The tendon of the_____ group of muscles contains a very large sesamoid bone, the_____
15. The innominate bones are each composed of three original parts, the _____, the_____ and the_____
16. The sacroiliac joints are most basically classified as _____ joints
17. The pelvic muscles may be grouped into the pelvic floor muscles, the_____ muscles, and those inserting on the _____
18. In walking, the heel is lifted by action of the _____ and _____ muscles, which are located in the _____
19. The gluteus minimus muscles are important in the tilting of the _____ that brings the body over the fixed leg in walking
20. The ankle joint is a _____ joint which permits only_____-flexion and_____-flexion of the foot
21. The great tendon of the back of the ankle is called the_____ tendon. It represents the insertion of the_____ and_____ muscles
22. Ordinarily the knee joint is capable only of _____ action
23. The accessory cartilages of the knee joint, easily torn by twisting or by a blow, are called the _____
   _____

92

24. The neck of the femur cannot form a callus if broken because it lacks a _____
25. The hip joint is a _____-____-_____ type of synovial joint
26. The actual movement of the thigh is restricted and limited by the action of _____ muscles
27. The bone of the upper arm is called the_____ It articulates proximally with the_____ and distally with the _____ which in turn articulates with the _____
28. The scapula adds mobility to the arm and extra inches to the reach by acting as a fourth _____ of the arm. It articulates proximally only with the_____
29. The radius articulates with the ulna only at the _____ end
30. The metacarpophalangeal joints are classified as _____ joints, but permit about the same range of motion as do_____joints, except in the case of the metacarpophalangeal joint in the_____
31. The extrinsic muscles of the hand are located in the_____
32. Downward displacement and dislocation of the shoulder joint is ordinarily prevented by a group of muscles collectively called the_____ _____
33. The very large intervertebral articular portion of most vertebrae is known as the _____ They are separated from one another by cartilagenous_____ _____
34. The_____and the_____vertebrae, however, lack centra
35. The spinal cord passes through the _____ of each vertebra. This space is surrounded in part by the archlike_____
36. The vertebrae of the lower trunk (below the ribs) are known as the_____vertebrae.
37. Each true rib is connected to the_____ by an individual_____cartilage
38. Each abdominal muscle produces some characteristic body movement, and in addition, they all serve to support the_____
39. The atlantooccipital joint is a _____type of synovial joint

40. The atlantoaxial joint, however, is a _____ joint, the _____ process of the axis acting as a pin around which atlas can turn
41. The lower jaw consists of a single bone, the _____ The _____ forms the upper jaw
42. The _____ skull bone articulates with the atlas vertebra
43. The masseter muscle _____ the lower jaw
44. In chewing, the actions of the muscles on one side of the jaw are antagonized by muscles located on the _____ _____ of the _____
45. The jaw joint is called the _____ joint

ANSWERS TO PRE-TEST QUESTIONS

1. contraction (LO 1)   2. converge (LO 1)   3. origins--broader (LO 1)   4. a muscle that works together with other muscles (LO 2)   5. a rounded projection of a bone (LO 2)   6. an opening into some body passageway (LO 2)   7. (a) the main axis of the foot passes through the big toe (b) the little toe counterbalances the action of the big toe (LO 4)   8. arch (LO 4)   9. interosseous--fibula (LO 4)   10. first--resistance (LO 3)   11. third (LO 3)   12. force--travel distance (LO 3)   13. femur (LO 5)   14. quadriceps--patella (LO 5)   15. ilium--ischium--pubis (LO 7)   16. synovial (hinge is a barely acceptable answer, which we tried to avoid in the wording) (LO 7, 10)   17. trunk--femur (LO 7)   18. gastrocnemius--soleus--shank (LO 6)   19. pelvis (LO 6)   20. hinge--dorsi--plantar (LO 4,6)   21. achilles--soleus--gastrocnemius (LO 4)   22. hinge (LO 4)   23. menisci (LO 4)   24. periosteum (LO 4)   25. ball-and-socket (LO 4)   26. fixator (LO 6)   27. humerus--scapula---ulna--radius (LO 8)   28. segment--clavicle (LO 8)   29. proximal (LO 8)   30. saddle--hinge---thumb (LO 8)   31. forearm (LO 9)   32. rotator cuff (LO 9)   33. centrum--intervertebral discs (LO 10) 34. atlas--axis (LO 10)   35. vertebral foramen--lamina (LO 10)   36. lumbar (LO 10)   37. sternum--costal (LO 10)   38. viscera (LO 10)   39. hinge (LO 10)   40. pivot--odontoid (LO 10,11)   41. mandible--maxilla (LO 11)   42. occipital (LO 11)   43. raises (LO 12)   44. other (or contralateral)--side--jaw (LO 12)   45. temporomandibular (LO 12)

LO 1: Give a functional interpretation of muscle shapes, that is, relate the shapes of muscles to the way in which they function in the body.

The basic, and in a sense, the only, muscular action is contraction. But the force of contraction must be channeled and usually concentrated upon an insertion which is usually much narrower than the origin, and generally via a tendon upon which muscle fibers themselves are inserted at an angle.

1. In most muscles the fibers converge upon a more or less central_____
2. Such an arrangement often produces a tapering_____ or_____shape
3. The bulk of a muscle tends to lie_____to the part of the body that it moves
4. Muscular_____tend to be broad, and_____ narrow, producing the typical_____shape of the limbs

TEST YOURSELF
1. In the case of most tendons a. typically they receive muscle fibers along much of their length b. they occur principally in the intercostal muscles c. they permit a muscle to insert at a distance d. all of the preceding
2. Such a muscle shape as this is classed as

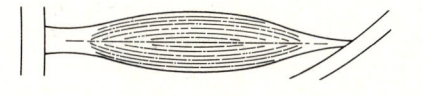

   a. penniform, b. triangular c. two-headed (bicipital)
   d. two-bellied e. fusiform

> *LO 2: Summarize the distinctive general terminology applied to muscles and bones in human anatomy.*

Make flash cards for the list of anatomic terms found in the textbook chapter 7. Write the term on one side and the definition on the other and study them as you would vocabulary words of a foreign language.

TEST YOURSELF

3. a prominent projection of a bone
4. a trench or shallow channel in a bone
5. a muscle that produces a particular action
6. an opening into some body passageway
7. a muscle which, with others, produces a combined action

a. agonist
b. meatus
c. eminence
d. synergist
e. fossa

> *LO 3: Identify the class of a given lever and be able to assign the major moveable body parts to an appropriate lever class.*

Levers are classified according to the relative position of the fulcrum which divides a lever into a force and a resistance arm. In a first class lever the force arm is long and in a third class lever it is short. In the case of a third class lever (to which class most of our moveable body parts belong, especially the limbs) a small muscular contraction produces a large movement of the extreme or distal part of the body portion but at a sacrifice of power which necessitates the provision of large, proximal muscle masses.

1. What is a
   a. fulcrum_____
   b. force arm_____
   c. resistance arm_____
2. Define a. first-class lever_____
   _____b. third-class lever_____
3. a. To what class do most of the body's levers belong
   b. Why_____
   _____

96

TEST YOURSELF
8. First-class levers a. are the commonest type in the body b. have a long resistance arm c. sacrifice power for distance traversed d. have a fulcrum located farthest from the force e. THREE of the preceding

> LO 4: List and describe the principal bones and joints of the thigh, leg and foot, classify them by type, and describe their range of motions

1. Complete the chart below the diagram.

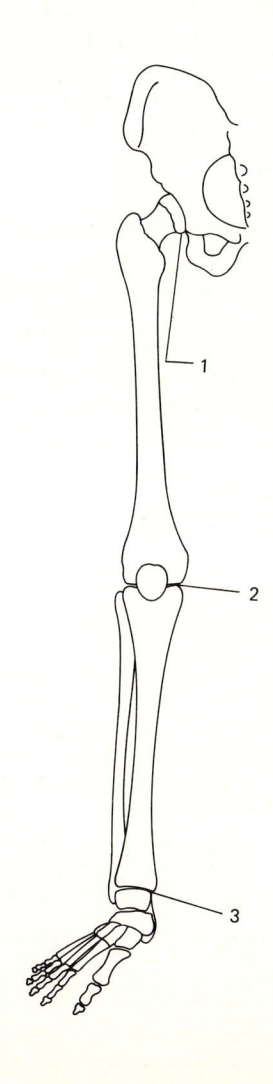

JOINT	NAME	TYPE	MOTION PERMITTED
1			
2			
3			

2. COMPLETE THIS CHART

NAME OF BONE	BRIEF DESCRIPTION	ARTICULATES WITH
Femur		
Tibia		
Fibula		
Tarsals (as a group)		
Metatarsals		

TEST YOURSELF

9. articulates with pelvis
10. rotates only when flexed
11. ball-and-socket joint
12. larger of the two shank bones
13. attached to Achilles tendon

a. hip joint
b. knee joint
c. tibia
d. femur
e. calcaneus

LO 5: Be able to arrange the thigh and leg muscles into functional groups; that is, flexors, extensors, adductors, abductors and rotators.

COMPLETE THIS CHART:

JOINT	FLEXORS	EXTENSORS	ABDUCTORS	ADDUCTORS	ROTATORS	FIXATORS
ANKLE						
KNEE						
HIP						

TEST YOURSELF

In the following questions, use the supplied choices.
in each case, choose the MAIN action of the muscle.

14. peroneus longus	a. flexor
15. biceps femoris	b. extensor
16. sartorius	c. abductor
17. gluteus maximus	d. adductor
18. gluteus minimus	e. rotator or
19. iliacus	fixator
20. quadriceps	

21. adductor of thigh	a. adductor brevis
22. dorsiflexor of foot	b. gracillis
23. plantar flexor of foot	c. gastrocnemius
	d. tibialis anterior
	e. (a) and (b)

LO 6: List agonists (using the names of supplied muscles) in typical leg movements such as standing erect from a squatting position, kicking, etc., and be able to describe the action of involved leg and trunk muscles in walking.

NOTE: There are an almost unlimited number of possible leg motions, and a large number of muscles in the leg which in various combinations produce these motions. Much can be done if one knows the actions of the main leg muscles very well indeed. By freely consulting the appropriate tables in the course of this exercise, you will begin to learn them.

1. FORWARD KICK
   a. muscles which move the shank, acting across the knee joint

   b. muscles which move the thigh (in this case flexing, acting across the hip joint

    c. lateral and medial muscles of the thigh (which by isometric contraction act to fix the thigh in this function)

2. SCISSORS KICK (employed in swimming the breast stroke)
    a. outward kick agonists

    b. inward kick agonists

    c. muscles employed in keeping the shank straight

3. WALKING (DOWN-step)
    a. describe action of gastrocnemius and soleus

    b. action of peroneus

    c. list extensors of shank (are these located on anterior or posterior aspect?)

    d. list extensors of femur. Anterior or posterior?

e. actions of the gluteus muscles

f. muscles involved in tilting and rotating the pelvis (other than glutei)

g. By careful study of subsequent tables, can you add anything not mentioned in the text regarding the action of trunk, spine and superior pelvic muscles in walking?

h. Briefly summarize muscle action as you would imagine it in the UP-step

LO 7: Describe the components, the shape and the function of the pelvis, and arrange the pelvic muscles into functional groups.

The pelvis (formally called the innominate bone) is a basin-shaped structure composed of two halves each of which articulates with the sacrum by a complex synovial joint, the sacroiliac. The two innominates are each composed of the ilium, ischium and pubis. These last are embryonically separate but fused in the adult. Classified by function, the major pelvic muscles of the thigh are as follows: adductors-- A. longus, brevis and magnus; abductors-- tensor fascia lata, gluteus minimus and piriformis; extensors--gluteus maximus, quadriceps, semitendinosis and sartorius (most adductors are also involved to some extent in extension); rotators--semimenbranosus, gracilis, semitendinosis, and sartorius; flexors--iliacus and psoas major (also serve to tilt the pelvis in walking, in conjunction with the glutei medius and minimus). The trunk muscles of the pelvis are: external oblique, rectus abdominus, psoas *minor, transversus abdominus, quadratus lumborum, sacrospinalis (the preceding mostly produce motions of the trunk itself), and the accessory limb muscles tensor fascia lata and latissimus dorsi. See tables 7-2 through 7-4, (and please note that very few muscles can be said to have a single, simple action. To get a simple movement of the thigh, for example, a number of muscles must be used simultaneously in such a way that the components of their actions irrelevent to that particular function mutually cancel).*

1. Describe the shape of the pelvis, or sketch it, indicating the contributions of the ilium, ischium and pubis

2. How is its shape related to its function, especially to its muscular attachments_____
   _____
3. In what pelvic structure can one find all three embryonic components_____
4. What are the two main divisions of the pelvic musculature a._____
   b._____
5. List muscles as follows, giving origin and insertion in each case:
   a. thigh adductors

b. thigh abductors

c. thigh flexors

d. thigh extensors

e. pelvic muscles moving the trunk

f. trunk-limb muscles

6. What muscles would be used to in-rotate or out-rotate the thigh?

TEST YOURSELF

24. Which of the following is NOT a part of the innominate bone   a. acetabulum   b. ilium   c. sacrum   d. ischium   e. pubis

25. quadriceps
26. iliacus
27. gluteus minimus
28. adductor magnus
29. sartorius

a. abductor
b. adductor
c. flexor (of thigh)
d. extensor
e. rotator

LO 8: List and describe the principal bones and joints of the arm, hand and shoulder; classify them as to type, describe their range of motions and summarize how each of them contributes to the use of the arm and hand.

The arm <u>functionally</u> consists of the scapula, humerus and ulna. The hand, whose basic construction is similar to that of the foot, consists of carpals, metacarpals and phalanges. The very loose shoulder (humeroscapular) joint is nearly universal in its freedom of motion. The elbow joint is, like the knee, basically a hinge joint but because of the proximal radioulnar joint the forearm is capable of extensive rotation referred to by the special terms pronation and supination. The wrist, or radiocarpal joint is also a hinge joint. The carpals, metacarpals and phalanges are arranged like the corresponding foot bones but are much more mobile. The thumb carpometacarpal joint is particularly mobile, being nearly universal in its action, permitting opposition of the thumb and fingers. The proximal end of the forelimb articulates to the trunk via the joint between the scapula and the clavicle. The action of the scapula adds inches to the reach and permits one to elevate the upper arm above shoulder level so that it is possible to reach far above the head.

1. The single proximal articulation of the scapula is with the_____
2. The ulna articulates with the_____and the _____
3. The radius articulates with the_____and the _____
4. In pronation and supination, the_____ rotates around the_____
5. The head of the_____articulates with the scapula at the_____ _____
6. The lengthwise ridge of the_____surface of the scapula is called its _____
7. The_____permits one to elevate the upper arm above the level of the shoulder

8. Label this illustration:

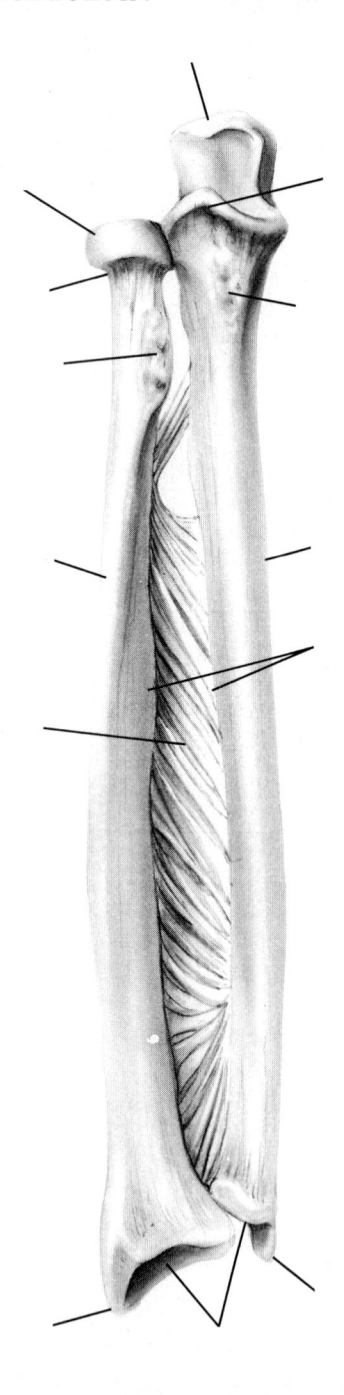

106

9. Label this illustration:

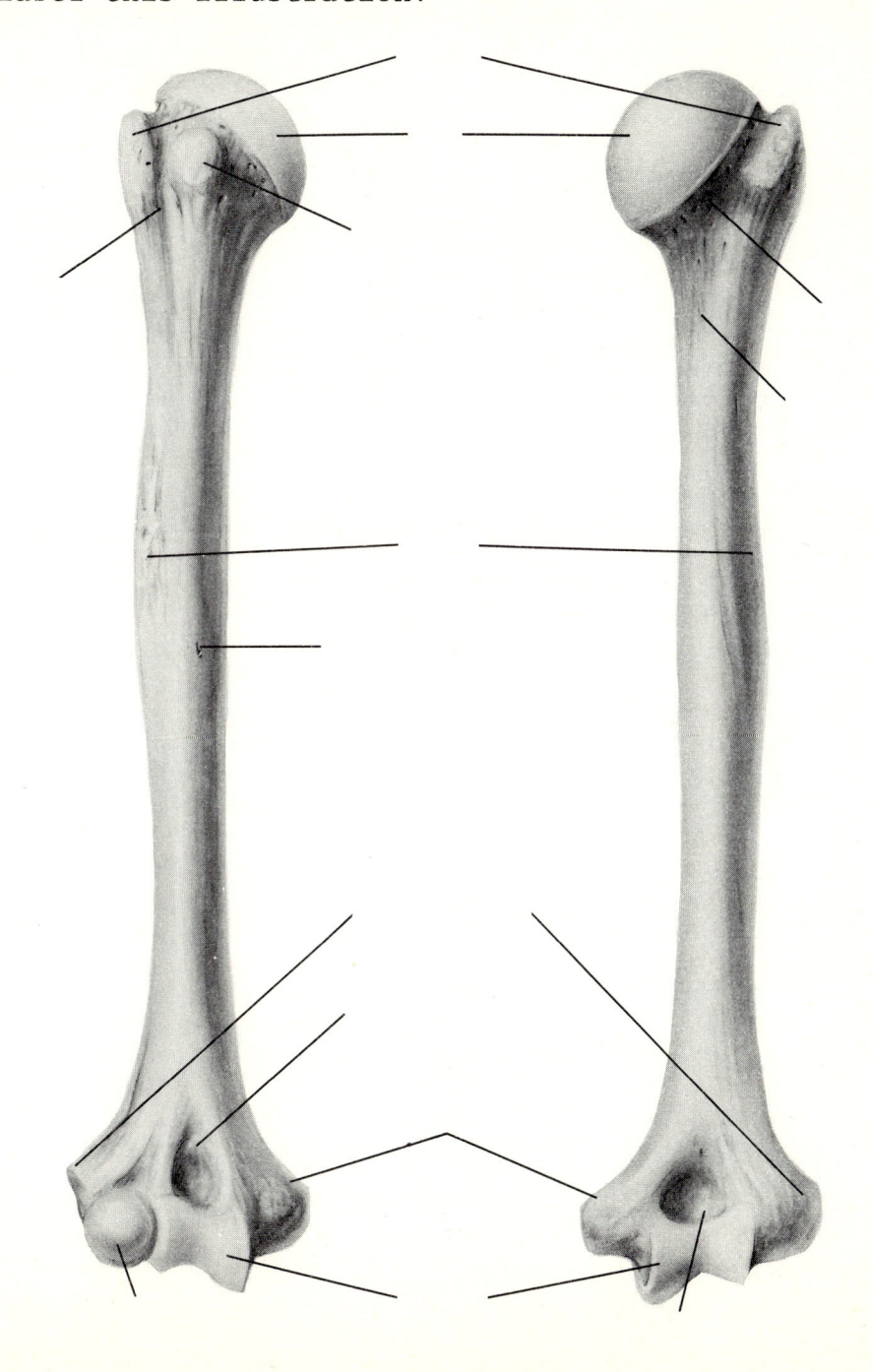

TEST YOURSELF (Some questions may have more than one
                correct answer)
30. possesses a raised ridge, the <u>spine</u>   a. humerus
31. possesses an interosseous membrane   b. scapula
32. articulates with the humerus   c. radius
33. does not move in supination   d. carpals
34. articulates with the clavicle   e. ulna
35. form wrist joint
36. upper arm bone

> LO 9: a) Group the muscles of the shoulder, arm and hand into functional categories, i.e., flexors, extensors, etc. b) group the trunk muscles into functional groups.

LO 9a has been done for you in Table 7-7 of the textbook, and there seems little point in repeating it here. The muscles of the trunk are divided first of all into anterior and posterior groups. The posterior division for the most part tends to keep the back erect and by unilateral or asymmetrical contraction to flex the spine--except for the suboccipital group which are neck muscles.

Most of the anterior trunk muscles are probably involved in respiration at least as accessories employed in extremity, but the main muscles of respiration are the diaphragm (inspiration) and the intercostals (mostly inspiration). Expiration is a mainly passive process, assisted when necessary by the sheetlike abdominal muscles. The abdominal muscles are also important in the containment of the abdominal viscera. Such muscles as the pectoralises and the rhomboidei are basically shoulder girdle muscles contributing to the mobility of the scapula or helping to fix it in position.

It should be noted also that the abdominal muscles brace the spine to a surprising extent. When one lifts a heavy load, the abdominal muscles are automatically tensed, producing a force transmitted approximately horizontally through the viscera to the spine which otherwise has a tendency to forward dislocation, especially at the lumbosacral location. Exercises strengthening the abdominal muscles therefor tend to prevent low back pain. From 50 to 80 percent of those who lose time from work because of spinal problems have abdominal muscles judged to be abnormally weak.

NOTE: *The following questions are only a sample of those that could be asked. You should make up some of your own to fit the approach chosen by your instructor. We also suggest that you exchange these with your fellow students.*

1. Which muscles pronate the arm _____
   _____

2. Which muscles supinate the arm _____
   _____

3. Why do none of the above muscles insert upon the humerus _____

4. Which arm muscle(s) act(s) across TWO joints _____

5. If the wrist is basically a hinge joint, how can the hand be adducted or abducted _____

6. Upon which bone do the adductors of the hand mostly originate _____ The abductors _____

7. Does either group tend to have a consistent flexor or extensor function _____

8. What is the difference between pronation-supination and just plain rotation of the arm _____

9. List the muscles of the rotator cuff _____

10. Other than rotation, what is their function _____
    _____

11. List the muscles of the back involved in maintaining the erect posture of the thoracic and lumbar spine
    _____
    _____
    _____

12. List the sheetlike abdominal muscles_____

13. Arrange the above into antagonistic groups (you will have to take origins, insertions, and muscle fiber directional arrangement into account in order to figure this out)_____
_____
_____

14. What relationship exists between the oblique abdominal muscles and the intercostals_____

15. Consider the basic arm actions: flexion, extension, rotation and supination-pronation, adduction, abduction and circumduction. Which one of these could serve as the basis of an exercise designed to strengthen all the upper arm and shoulder muscles? Why_____
_____
_____

TEST YOURSELF

37. flexor carpi radialis
38. flexor carpi ulnaris
39. subscapularis
40. biceps brachii
41. palmaris longis

a. adductor
b. abductor
c. pronator
d. supinator
e. rotator (of humerus)

42. Which of the following are antagonists   a. extensor carpi radialis longus--flexor carpi radialis
    b. extensor carpi radialis--extensor carpi ulnaris
    c. flexor carpi radialis--biceps brachii
    d. deltoid--pectoralis major   e. all of the above

43. semispinalis capitis
44. rectus capitus post. major
45. external oblique
46. scalenes
47. rhomboideus

a. extends, rotates head
b. erects spine
c. contains abdominal viscera
d. draws scapula upward
e. flexes spine

LO 10: *Describe the axial skeleton (with the exception of the skull), laying emphasis upon the differentiated regions of the vertebral column and the structure of typical individual vertebrae.*

The axial skeleton is here considered to consist of the spinal column, the ribs, the sternum and the skull. The flexible vertebral column is kept firm by muscular action and is sometimes flexed, laterally as well as anteroposteriorly, by those same muscles. The vertebral column is composed of the cervical, thoracic, lumbar and sacral portions, plus the coccyx. Each thoracic vertebra articulates with its neighbors in two ways: via the centrum and its associated intervertebral disc, and also by the synovial inferior and superior apophyseal joints. Each thoracic vertebra also articulates with one of the heads of a rib at each end of the vertebra, and via a middle transverse process also.

1. With what bones, other than vertebrae, do some of the vertebrae articulate? Which regions of the vertebral column are involved in each case
   a._____
   b._____
   c._____
2. How do the atlas and axis differ from more typical vertebrae   a)_____

   b) from one another_____

3. What is the function of the odontoid process_____

4. Describe the synovial articulation of the thoracic vertebrae: a) with one another _____

   b) with the ribs_____

5. What is the function of an intervertebral disc____

6. Give two kinds of intervertebral disc lesion, and describe each: a)_____

   b)_____

8. What are the three parts of the sternum
   (a) _____
   (b) _____
   (c) _____
9. What kinds of ribs are usually distinguished
   (a) _____
   (b) _____
   (c) _____

TEST YOURSELF

48. muscle attachment
49. region of intervertebral disc attachment
50. comprise the neural canal's roof
51. intervertebral synovial articulation

a. lamina
b. centrum
c. apophysis
d. vertebral foramen
e. spinous process

52. The annulus fibrosus is a. the outer part of the intervertebral disc b. the pulpy part of the apophysis c. one of the accessory cartilages of the knee joint d. the cartilage of the sacroiliac joint e. an intervertebral ligament often broken or strained in football

53. The atlas vertebra a. lacks a centrum b. possesses an odontoid process c. articulates with the occipital bone d. all of the preceding e. two of the preceding

> LO 11: List and briefly describe the bones of the skull and be able to identify each on an unlabelled diagram.

This learning objective has been done for you in table 7-9 of the textbook, so no summary of it is here attempted. But we do suggest very careful study of the table along with <u>all relevent illustrations</u>.

Label the diagrams on the following pages (113-4-5-6)

7. Label the following diagram:

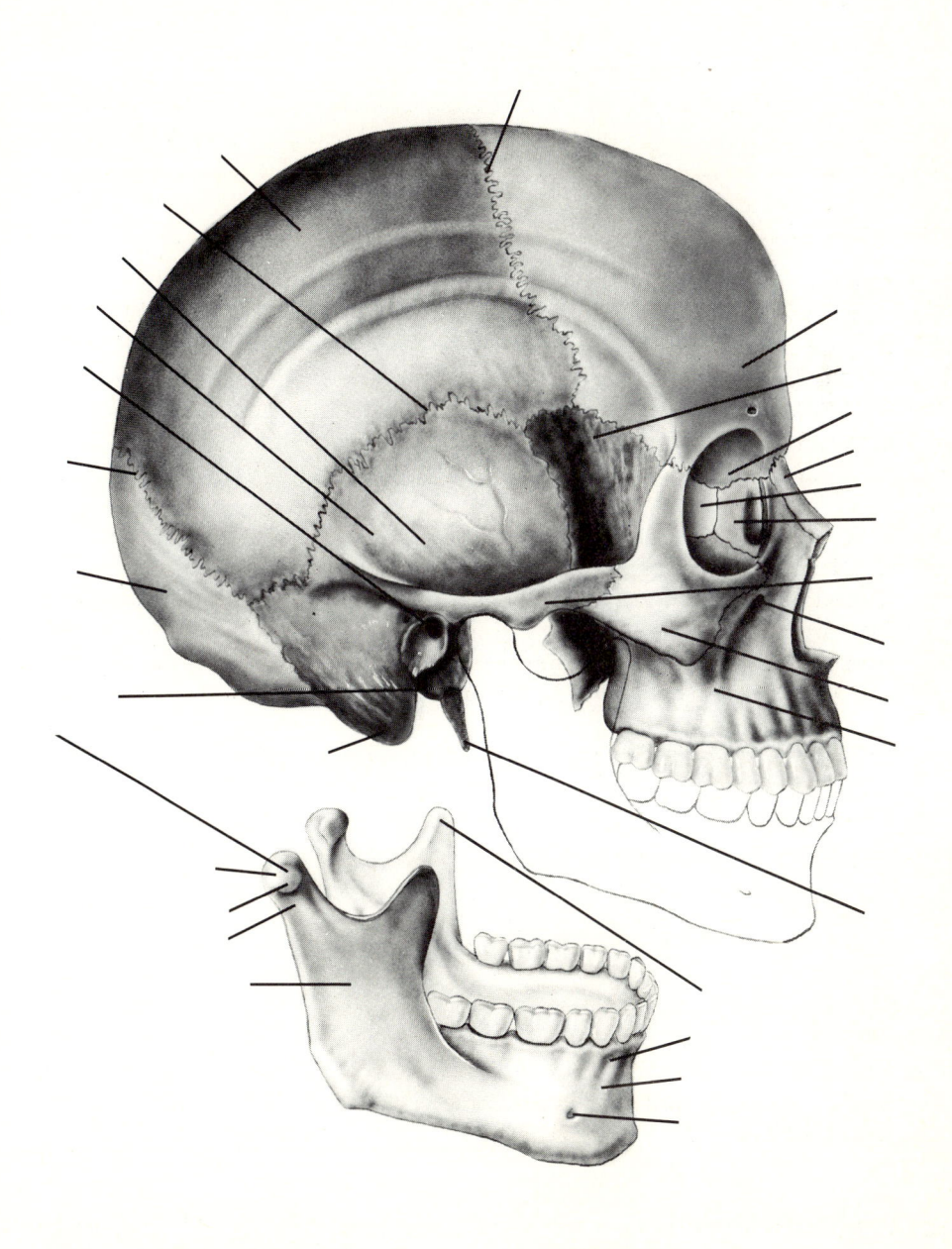

114

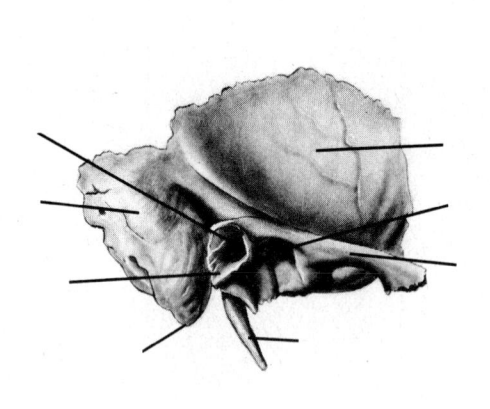

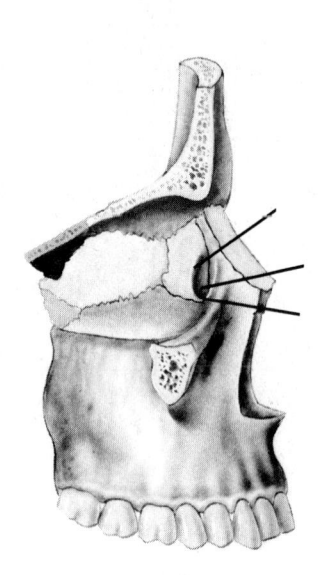

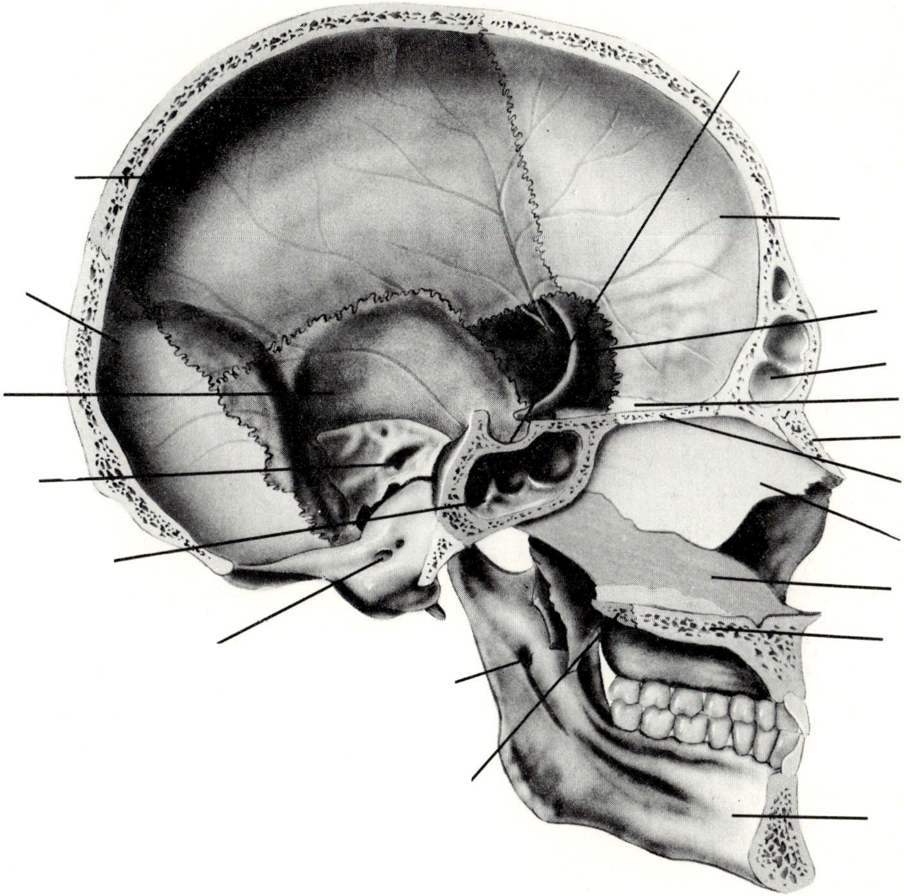

116

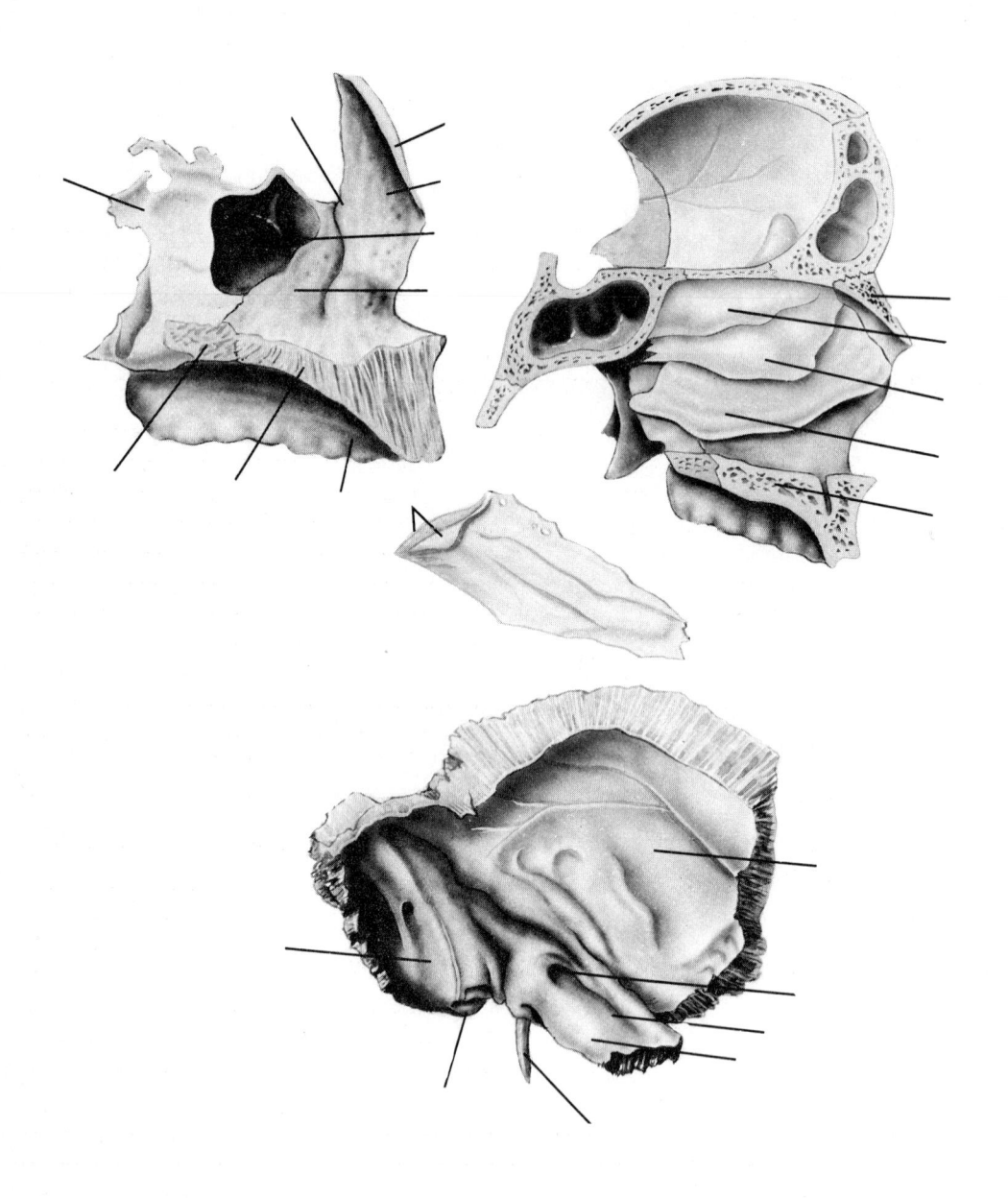

TEST YOURSELF

54. main bone of the forehead
55. articulates with lower jaw
56. contains the ear
57. bears depression for pituitary gland
58. walls of braincase(lateral)
59. cheekbone

a. parietal
b. temporal
c. frontal
d. ethmoid
e. malar

> LO 12: List agonist muscles in such representative motions of the head and neck as nodding, shaking and chewing; and in each case describe joint action.

1. Summarize the combined actions of the atlanto-occipital and atlanto-axial joints _____
   _____
   _____

2. Do the same for the TMJ _____
   _____
   _____

3. Which muscles are active in mastication _____
   _____

4. Which muscles are <u>antagonists</u> in closing the mouth ___
   _____

5. Which muscles are antagonists in moving the jaw from side to side as in chewing _____
   _____

6. List the major muscles agonistic in raising the head (extending the neck) _____
   _____

7. Now give their antagonists _____
   _____
   _____
   _____

8. Which of the preceding sets of muscles, No. 6_____ or No. 7_____, would be contracted when one slowly lowers his chin to touch his chest? (this is not as easy a question as perhaps it might seem)

9. What is the principle of muscle action involved in shaking the head_____
_____

10. As an optional project, try to list the muscles involved in adducting and abducting the head on a separate sheet

11. Do the same, listing them in order of their contraction, for circumducting the head

TEST YOURSELF

60. atlantoaxial
61. atlantooccipital
62. temperomandibular
63. frontoparietal

a. hinge joint
b. pivot joint
c. sliding joint
d. ball-and-socket joint
e. none of the above

64. The pterygoideus medialis muscle, in side-to-side chewing action, is antagonized by the a. masseter of the same (ipsilateral) side b. digastric c. contralateral pterygoideus medialis and similar acting muscles d. trapezius e. ipsilateral temporalis

65. Which of the following flexes the neck a. trapezius b. sternocleidomastoid c. splenius capitis d. none of the preceding

LO 13: Be able to describe all muscles specified by your instructor with respect to location, origin, insertion, action, allies and antagonists, and typical function.

In our opinion there is only one good way to fulfil this objective, if it is assigned by your instructor, and that is to make flash cards. Use ordinary index cards. Place the name of the muscle on one side, and on the other give origin, action, allies (synergists), antagonists, and typical function. By this last we mean a set of natural body motions in which it assists other muscles that would be expected to act along with it. Thus the quadriceps is, among other things, an extensor of the shank. It would be active both in squatting and in raising oneself up from such a position (as in doing deep knee bend exercises).

Of course your instructor may require either more or less than this, or may delete or add muscles to the lists we have provided. We have no way of predicting his or her exact objectives in advance which is part of the reason we do not give you more material in connection with this LO. But you can easily use your flash cards to obtain the same effects as a quiz. Just turn them over to find the answer. By arranging them in functional groups, or by sorting them out, you can learn all the muscles which have a similar action. This will help you to fulfill the "typical function" part of this LO. Best of all, learning will be accomplished not only by using these cards, but (on a subconscious level) even when you prepare them.

ANSWERS TO ALL TEST-YOURSELF QUESTIONS

*1.c   2.e   3.c   4.e   5.a   6.b   7.d   8.d   9.d
10.b   11.a   12.c   13.e   14.a   15.b,d 16.a   17.b
18.c   19.e   20.b   21.e   22.d   23.c   24.c   25.d
26.c   27.a   28.b   29.c,e 30.b   31.c,e 32.e
33.e   34.b   35.c,d 36.a   37.b   38.a   39.e
40.d,b 41.c   42.e   (Note: of course in some functions some of these pairs may not be antagonistic. Antagonism depends not only upon action in the usual sense but also upon specific function in a particular body movement)   43.b   44.a   45.c   46.e   47.d
48.e   49.b   50.d   51.c   52.a   53.e   54.c   55.b
56.b   57.d   58.a   59.e   60.b   61.a   62.c   63.e
64.c   65.d*

POST-TEST

1. (LO 1) A muscle can contract up to about _____ % of its relaxed length   a. 75   b. 50   c. 25   d. 5

2. (LO 2) A muscle that acts, along with others, so that the sum of their action produces a particular body movement, is best called a(n)   a. agonist   b. antagonist   c. synergist   d. fixator   e. prime mover

3. (LO 2) An opening into a body passage, not necessarily bony, is called a(n)   a. sulcus   b. epicondyle   c. sinus   d. meatus   e. trochanter

4. (LO 3) In a third class lever   a. the force arm is long   b. the resistance arm is long   c. speed is sacrificed to obtain power   d. two of the preceding   e. none of the preceding

5. (LO 4) Which of the following is NOT a hinge joint   a. talotibial   b. tibiofemoral   c. hip   d. all of the above are, in fact, hinge joints   e. none of the above are hinge joints

6. (LO 4) The shank is capable of some rotation (at the knee joint)   a. when extended   b. when flexed   c. when the thigh is abducted   d. at no time   e. at all times

7. (LO 5, Table 7-2) Which of these is both a thigh flexor and a shank extender   a. semitendinosus   b. biceps femoris   c. gracilis   d. gastrocnemius   e. quadriceps

8. (LO 6, Table 7-3) Which of the following tilts the pelvis sideways or rotates it at the end of a stride   a. gluteus maximus   b. adductor longus   c. biceps femoris   d. iliacus   e. more than one of the preceding

9. (LO 7) The sacrum articulates with the   a. femur   b. ilium   c. ischium   d. pubis   e. (b),(c) and (d) all participate in the joint

10. (LO 7 and 9b, Table 7-4) Which of the following muscles would elevate the trunk in a sitting-up exercise a. tensor fascia lata b. rectus abdominus c. quadriceps d. latissimus dorsi e. (b) and (c)

11. (LO 8) Which of the following arm joints is much more mobile than its equivalent in the leg a. wrist b. finger (metacarpophalangeal) c. finger (interphalangeal) elbow e. shoulder

12. (LO 8) Which of the following though not a hinge joint, has the action of a hinge joint a. scapulohumeral b. humeroulnar c. wrist d. metacarpophalangeal of finger e. (b),(c) and (d)

13. (LO 9a, Table 7-7) Which of the following is not a forearm muscle a. flexor digitorum profundis b. extensor pollicis c. opponens pollicis d. extensor digitorum communis e. (b) and (c)

14. (LO 9a, Table 7-6) Which of the following acts across two joints a. biceps brachii b. deltoid c. latissimus dorsi d. pronator teres e. brachialis

15. (LO 9a, Table 7-6) What is the action of the muscle that you chose in question 14 a. pronation b. extension c. flexion d. supination e. (c) and (d)

16. (LO 9a, Table 7-7) Which of the following is NOT part of the "rotator cuff" a. supraspinatus b. infraspinatus c. deltoid d. teres minor e. subscapularis

17. (LO 9b, Table 7-8) Which of the following muscles are used in quiet inspiration a. diaphragm b. internal intercostals c. external intercostals d. all of the preceding e. (a) and (b) but not (c) which is solely expiratory

121

18. (LO 9b, Table 7-8) Which of the following adducts the arm a. erector spinae group b. pectoralis major c. transversus abdominus d. semispinalis group e. external oblique

19. (LO 10) Which of the following vertebrae differs most radically from any and all of the others a. the superiormost cervical b. the inferiormost cervical c. the third thoracic d. the superiormost lumbar e. the inferiormost lumbar

20. (LO 10) Which of the following joints is NOT synovial a. rib-vertebra(costovertebral) b. apophyseal intervertebral c. intercentral intervertebral d. sacroiliac e. (c) and (d)

21. (LO 11) Which of these is the most inferior a. frontal b. parietal c. malar d. nasal e. sphenoid

22. (LO 12, Table 7-10) Which of the following is a muscle used in chewing a. temporalis b. splenius capitis c. sternocleidomastoid d. orbicularis oculi e. epicranius

ANSWERS TO POST-TEST QUESTIONS

1.b	2.c	3.d	4.b	5.c	6.b	7.e	8.d	9.b
10.e	11.d	12.d	13.c	14.a	15.e	16.c	17.d	
18.b	19.a	20.c	21.e	22.a				

8 THE NERVOUS SYSTEM
Basic Organization and Neurophysiology

PRETEST

1. For the sake of study the nervous system may be divided into two broad divisions the _____ _____ system and the _____ _____ system
2. The CNS consists of _____ and _____
3. The PNS may be divided into a somatic portion and an _____ system
4. Phagocytic cells found within the nervous tissue are called _____
5. The nucleus of a neuron is found within its _____
6. The _____ functions to transmit impulses away from the cell body and towards an effector
7. The _____ _____ insulates the axon
8. The _____ sheath is essential to regeneration of an injured neuron
9. _____ neurons transmit messages from the CNS to the muscles and glands
10. A cluster of nerve cell bodies located outside of the CNS is called a _____
11. The first step in neural response is _____
12. In a polysynaptic reflex pathway an impulse passes from a sensory neuron to an _____ neuron
13. Many reflex actions occur without the participation of the _____
14. Receptors convert various forms of energy into _____ impulses

15. In a resting neuron the inner surface of the cell membrane is _____ charged as compared to the interstitial fluid outside
16. A stimulus that increases the permeability of the neuron to sodium may result in propagation of a _____ _____
17. The number of neurons stimulated and the frequency of response of the neurons influence the _____ of a sensation
18. In saltatory conduction the impulse jumps along the axon from one _____ _____ to another
19. When insufficient numbers of calcium ions are present neurons fire _____ easily
    more or less
20. Transmission of neural impulses across synapses depends upon release of _____ _____
21. The type of transmitter substance released at myoneural junctions is _____
22. Large neurons (i.e., those with the greatest diameter) transmit impulses _____
    fastest, slowest
23. The process of adding excitatory postsynaptic potentials is known as _____
24. A single neuron may synapse with hundreds of postsynaptic neurons. This is known as _____
25. In a _____ circuit neuron branches turn back upon themselves stimulating prolonged activity

ANSWERS TO THE PRETEST QUESTIONS

1. central nervous, peripheral nervous (LO 1) 2. brain spinal cord (LO 1) 3. autonomic (LO 1) 4. microglia (LO 2) 5. cell body (LO 3) 6. axon (LO -) 7. myelin sheath (LO 3) 8. cellular (LO 4) 9. motor (or efferent) (LO 5) 10. ganglion (LO 6) 11. reception (LO 7) 12. association (LO 8) 13. brain (LO 9) 14. neural (LO 10) 15. negatively (LO 11) 16. neural impulse (or action potential) (LO 12) 17. intensity (LO 13) 18. node of Ranvier (LO 14) 19. more (LO 15) 20. transmitter substance (LO 16) 21. acetylcholine (LO 17) 22. fastest (LO 18) 23. summation (LO 19) 24. divergence (LO 20) 25. reverberating (LO 21)

LO 1: List the divisions of the nervous system, and identify their main functions.

Two main divisions of the nervous system are the central nervous system (CNS) and peripheral nervous system (PNS). The CNS includes brain and spinal cord; the PNS which consists of nerves and ganglia is divided into somatic and autonomic portions.

1. To which division does each of the following belong;
   a. brain_____ b. spinal nerve_____
   c. ganglion_____
2. What is the job of the somatic portion of the PNS _____
3. What are the two types of efferent pathways in the autonomic system _____
_____

TEST YOURSELF

1. The CNS consists of the a. brain b. spinal nerves c. sympathetic nerves d. autonomic system e. answers (a) and (b) are correct

2. The parasympathetic nerves are part of the a. CNS b. autonomic system c. somatic system d. sympathetic system e. brain

LO 2: Give the functions of the glial cells.

Glial cells play supportive roles in the nervous system. Oligodendrocytes form sheaths around neurons in the CNS and may assist in neuron nutrition. Microglia are phagocytes and ependymal cells line the cavities of the brain and spinal cord. Astrocytes may provide some structural support and play a role in the blood-brain barrier.

1. What is the general function of glial cells _____
2. Give the specific functions of a. microglial cells _____
   b. ependymal cells _____
_____

TEST YOURSELF

3. Following a stroke some injured cells within the brain die. The dead cells are phagocytized by a. astrocytes b. neurons c. microglial cells d. ependymal cells e. oligodendrocytes

4. Sheaths are formed about CNS neurons by a. ependymal cells, b. neurons from the PNS c. Schwann cells d. oligodendrocytes e. microglial cells

> LO 3: Draw or label on a given drawing a typical neuron, label its parts, and give the functions of each, including myelin and cellular sheaths.

In a typical neuron the dendrites and cell body receive and integrate neural messages while the axon conducts impulses and releases transmitter substance from its synaptic knobs. The myelin sheath insulates the axon and the cellular sheath functions in regeneration. Perform the exercise on the next page.

TEST YOURSELF

5. Synaptic knobs are found a. in the myelin sheath b. in the cellular sheath c. at the dendrite ends d. in the cell body e. at the axon endings

6. The part of the neuron that contains most of the cytoplasm as well as the nucleus is the a. axon b. dendrite c. cell body e. node of Ranvier e. cellular sheath

> LO 4: Describe the process of nerve regeneration.

When a neuron is severed the cut portion separated from the cell body disintegrates. The remaining axon ends send sprouts out into the cellular sheath which remains and serves as a guide for the growing fiber.

What happens if an axon sprout grows through a neighboring cellular sheath that ends in the wrong location

Label the diagram below, and indicate the function of each structure by writing as few key words as you can next to each label.

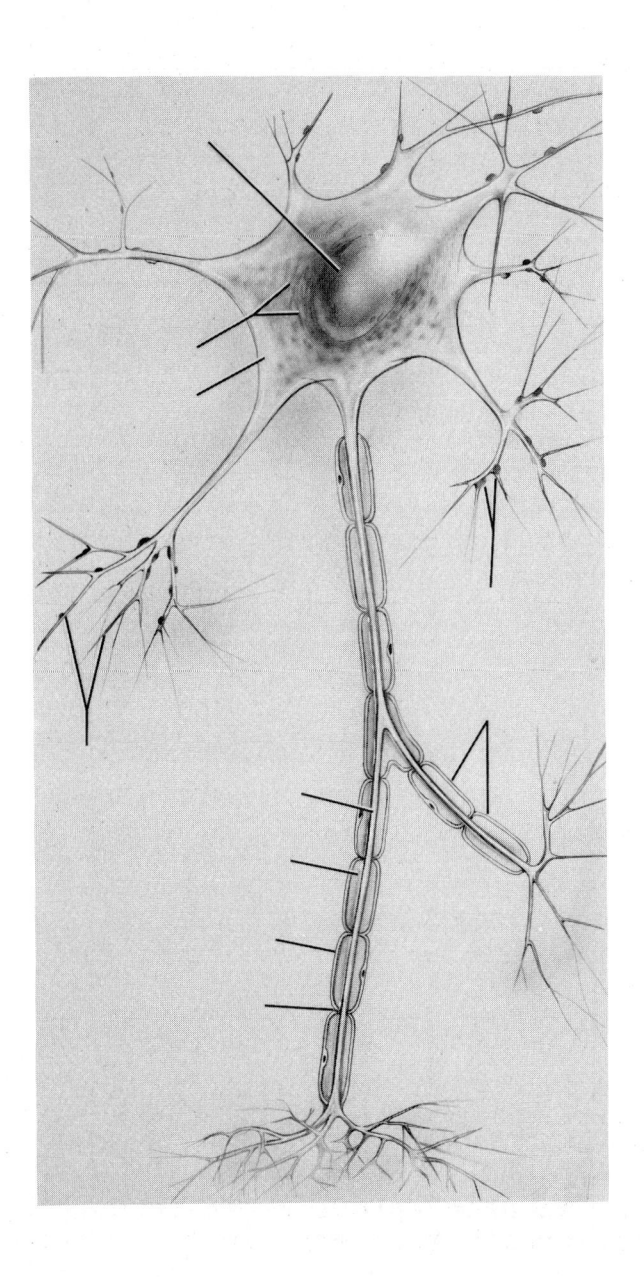

TEST YOURSELF

7. In nerve regeneration the cellular sheath a. acts as a guide for the new axon sprout b. proliferates new nerve cells by mitosis c. disintegrates if separated from the cell body d. may be so damaged as to cause multiple sclerosis e. two of the preceding answers are correct

> LO 5: List 3 principal types of neurons based on function.

Sensory neurons transmit impulses from receptors to the CNS. Association neurons connect sensory neurons with motor neurons. Motor neurons transmit messages from the CNS to the effectors.

1. Where are association neurons located_____
2. Draw a diagram to show how these 3 types of neurons may link with one another in a functional sequence

TEST YOURSELF

8. Neurons that link sensory and motor neurons are called a. afferent neurons b. efferent neurons c. ganglia d. association neurons e. unipolar neurons

9. Association neurons are found a. within the CNS b. within muscles c. in the skin d. passing from skin to muscles e. passing from skin to the CNS

> LO 6: Distinguish between nerve and tract: Ganglion and nucleus.

Nerves and ganglia are part of the PNS. Tracts and nuclei are located within the CNS.

1. Define each of the following:
   a. ganglion_____
   b. nucleus_____
   c. nerve_____
   d. tract_____

TEST YOURSELF

10. A large number of nerve fibers enwrapped in connective tissue (epineurium) and located outside the CNS is referred to as a   a. ganglion   b. nerve   c. tract   d. nucleus   e. multipolar neuron

> LO 7: Briefly describe the four basic processes upon which all neural responses depend....reception, transmission, integration, and effection.

1. Imagine that a monster suddenly appears before you and you quickly turn and run. Analyze this response in terms of the 4 processes listed:
   a. reception_____
   b. transmission_____
   c. integration_____
   d. effection_____

TEST YOURSELF

11. In the scenario just given, the actual running away is an example of   a. reception   b. transmission   c. integration   d. effection   e. a combination of reception and transmission

12. Deciding whether or not to run would be part of
    a. reception   b. transmission   c. integration
    d. effection

> LO 8: Diagram (or label on a given diagram) a monosynaptic and a polysynaptic reflex pathway, identifying their essential components and indicating the direction of impulse transmission.

In a reflex pathway a receptor triggers impulses in sensory neurons which transmit them to the CNS. In the monosynaptic pathway the sensory neuron synapses with a motor neuron that transmits information to an appropriate effector. In the polysynaptic pathway an association neuron is interposed between sensory and motor neurons.

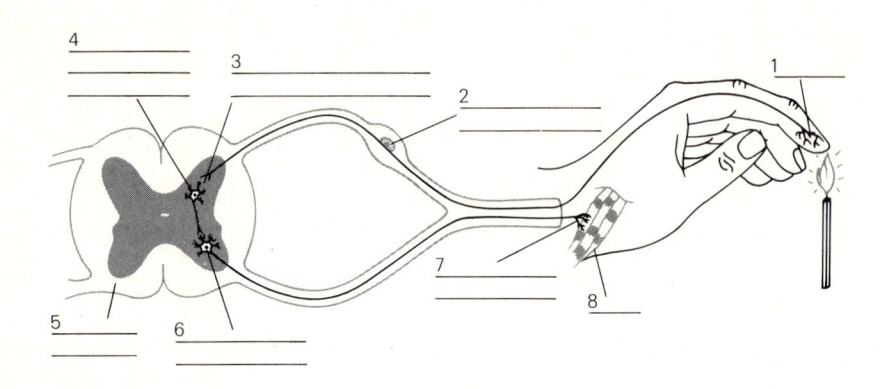

Fill in the table from the structures indicated in the preceding diagram.

STRUCTURE	FUNCTION
1.	
2.	
3.	
4.	
5.	
6.	
7.	

TEST YOURSELF

13. As shown in the diagram, reception takes place at the structure labeled   a. 1   b. 2   c. 7   d. 6   e. 5

14. In the diagram a synapse is labeled as number
    a. 1   b. 2   c. 3   d. 4   e. 5

15. The cell body of the association neuron is labeled
    a. 2   b. 3   c. 4   d. 5   e. 6

>   *LO 9: Describe an experiment to demonstrate that the brain is not essential to the activity of some reflexes.*

1. Recount the experiment cited in the textbook in which a frog was decapitated and its foot placed in an acid solution. What happened_____

2. What does the experiment illustrate_____
_____

TEST YOURSELF

16. In reflex pathways that do not involve the brain integration takes place in the   a. spinal cord   b. motor neurons   c. muscle   d. neck   e. receptors

>   *LO 10: Describe the general action of receptors.*

Receptors convert various forms of energy into neural impulses. Each type of receptor is specialized to react to one particular type of stimulus.

1. Do different types of receptors send different types of neural messages _____

2. How does the CNS "know" what an incoming message from a receptor says, for example, cold or pressure
_____
_____

TEST YOURSELF

17. If a neuron that normally transmits messages from a pressure receptor to the CNS is electrically stimulated, the person will perceive a. an electric shock b. a feeling of cold in the area served by that neuron c. a generalized feeling of pressure d. a feeling of pressure in the area served by that neuron e. no stimulus at all

> LO 11: Give the ionic basis of resting membrane potential.

Resting membrane potential is brought about by the action of the sodium pump which causes sodium ions to accumulate along the outside of the axon membrane and by the large number of negatively charged proteins and other molecules within the axon.

1. What is meant by the statement: The resting neuron is polarized_____

2. What is the sodium pump_____
   _____

3. What is a resting potential_____
   _____

TEST YOURSELF

18. Sodium ions accumulate outside the resting neuron membrane as a result of a. diffusion b. active transport c. osmosis d. polarity e. depolarization

> LO 12: Describe the propagation of an action potential.

An action potential is initiated when sufficient sodium is allowed to diffuse into the neuron to depolarize the axon to a critical threshold level.

132

1. What is meant by threshold level_____
2. What is a wave of depolarization_____
3. What is the all-or-none law_____
_____

TEST YOURSELF

19. An action potential is a a. process of polarization b. neural impulse c. permanent change in the polarization of the axon d. release of transmitter substance from the synaptic knob e. refractory period

20. A stimulus may result in propagation of a neural impulse if it a. decreases the permeability of the neuron to sodium b. increases the permeability of the neuron to sodium c. prevents a wave of depolarization d. violates the all-or-none law e. activates the sodium pump

> LO 13: Indicate two factors that affect intensity of sensation.

Intensity of sensation is affected by the number of neurons stimulated, and by their frequency of response.

1. How can we tell the difference between a little toothache and a big one? Explain the physiologic basis
_____

TEST YOURSELF

21. Mary has a small cut on her hand. Anne has a large cut. Anne's cut hurts more because a. more neurons are being stimulated by pain receptors b. due to a larger area of inflammation pain receptors continue to be stimulated c. stronger stimuli are resulting in more intense waves of depolarization along each axon affected d. answers (a),(b) and (c) are correct e. answers (a) and (b) only.

LO 14: Compare continuous with saltatory conduction.

In myelinated neurons the impulse jumps along from one node of Ranvier to the next. This is known as saltatory conduction.

1. How does saltatory conduction compare to transmission along an unmyelinated neuron _____
_____
_____

TEST YOURSELF

22. Saltatory conduction is a. characteristic of unmyelinated neurons b. requires more energy than continuous transmission c. is faster than continuous transmission d. requires the sodium pump to work harder e. two of the preceding answers are correct

LO 15: Give the effects of too much or too little calcium upon neural function.

Too much calcium makes neurons more difficult to fire, whereas too few calcium ions make neurons more irritable.

1. What effect does the presence of too few calcium ions have upon sodium balance_____

2. Why do muscles sometimes go into spasm when too few calcium ions are present_____
_____

TEST YOURSELF

23. When too few calcium ions are present the a. sodium tends to leak into the cell b. the resting potential is lowered c. the neuron fires more easily d. answers (a),(b) and (c) are correct e. answers (a) and (b) only are correct

LO 16: List and describe the steps in synaptic transmission. (Draw diagrams to support your description.)

Presynaptic axons release transmitter substance from their synaptic knobs. Molecules of transmitter substance combine with receptors upon the postsynaptic neurons, resulting in EPSPs and IPSPs.

1. Draw a diagram showing the release of transmitter substance from a presynaptic neuron and its uptake by a postsynaptic neuron

2. When do the vesicles release the transmitter substance_____
3. What happens to excess transmitter substance in the synaptic cleft_____
_____

TEST YOURSELF

24. When an impulse reaches the synaptic knobs at the end of an axon a. vesicles fuse with the cell membrane b. vesicles release their contents into the synaptic cleft c. specific receptors on the axon membrane of the presynaptic neuron take up the transmitter substance d. answers (a),(b) and (c) are correct e. answers (a) and (b) only are correct

25. Excess transmitter substance in the synaptic cleft may be a. reabsorbed into the synaptic vesicles b. inactivated by enzymes c. retained in the synaptic cleft for continuous use d. answers (a),(b) and (c) are correct e. answers (a) and (b) only are correct

LO 17: Identify the transmitter substances described in the chapter.

Acetylcholine, norepinephrine, dopamine, and serotonin are the transmitter substances which have been most thoroughly investigated. (Read about them in Table 8-1 of your textbook.)

1. What are cholinergic neurons_____

2. What are adrenergic neurons_____

3. Which transmitter substances are catecholamines____

4. Name the transmitter substance
   a. released at myoneural junctions_____
   b. thought to affect motor function_____
   c. that may play a role in sleep_____

TEST YOURSELF

26. The transmitter substance released by motor neurons at myoneural junctions is   a. acetylcholine   b. norepinephrine   c. dopamine   d. serotonin   e. adrenergic

27. Cholinergic neurons release   a. norepinephrine   b. epinephrine   c. acetylcholine   d. dopamine   e. serotonin

   LO 18: Identify factors that affect speed of transmission.

1. Describe type A fibers_____

2. List 3 factors that affect speed of transmission through a neural pathway
   a._____
   b._____
   c._____

TEST YOURSELF

28. Transmission is fastest   a. in large, myelinated axons   b. in pathways with the fewest number of syn-

apses   c. in Type C fibers   d. answers (a), (b) and (c) are correct   e. answers (a) and (b) only are correct

> LO 19: Describe how a postsynaptic neuron integrates incoming stimuli and "decides" whether or not to fire.

The surface of a postsynaptic neuron continually responds to EPSPs and IPSPs. When sufficient EPSPs are present to depolarize the membrane, an impulse is fired.

1. Define an EPSP in terms of sodium permeability _____
2. What is an IPSP_____
3. What is summation_____

TEST YOURSELF

29. If transmitter substance makes the membrane of a postsynaptic neuron more permeable to sodium   a. an excitatory postsynaptic potential (EPSP) has formed   b. the neuron is brought closer to firing   c. an inhibitory postsynaptic potential (IPSP) has formed   d. summation has occurred   e. answers (a) and (b) are correct

30. Summation occurs when   a. subliminal EPSPs are added together   b. transmitter substance combines with a receptor on the postsynaptic neuron   d. an IPSP reaches threshold level   d. the presynaptic neuron tabulates all of its vesicles   e. sodium ions are added together and then ejected from the neuron

> LO 20: Contrast convergence and divergence, and tell why each is important.

Convergence and divergence provide the opportunity for complex neural interactions at synapses, including facilitation and reverberation.

1. Describe and draw a diagram(on a separate sheet of paper) to illustrate:

   a. convergence_____
   b. divergence_____
                         _____

2. Define facilitation_____
   _____

TEST YOURSELF

31. An association neuron in the spinal cord receives information from sensory neurons, neurons coming from the brain, and neurons coming from different segments of the cord. This is an example of
a. divergence b. facilitation   c. convergence
d. summation e. an IPSP

> *LO 21: Describe a reverberating circuit. (Also describe afterdischarge.)*

A reverberating circuit is a neural pathway arranged so that branches turn back upon themselves stimulating prolonged activity.

1. On a separate sheet of paper draw a diagram to illustrate a reverberating circuit

2. On a separate sheet of paper draw a diagram to illustrate afterdischarge

TEST YOURSELF

32. Once activated the neurons in a reverberating circuit will continue to fire until a. they become fatigued b. they are stopped by inhibition c. the impulse reaches the end of the sequence d. the impulse is delivered to an adjoining pathway e. answers (a) and (b) are correct

Many new terms were used in this chapter. Have you kept a vocabulary list? How many new terms have you listed?

ANSWERS TO TEST YOURSELF QUESTIONS

1.a	2.b	3.c	4.d	5.e	6.c	7.a	8.d	9.a	10.b	11.d
12.c	13.a	14.c	15.c	16.a	17.d	18.b	19.b	20.b
21.e	22.c	23.d	24.e	25.e	26.a	27.c	28.e	29.e
30.a	31.c	32.e

POST TEST

1. (LO 1) The brain is part of the
   a. somatic system   b. peripheral nervous system
   c. central nervous system   d. autonomic system
   e. sympathetic system

2. (LO 1) Nerves that transmit information from the CNS to involuntary muscles belong to the   a. CNS   b. autonomic system, efferent portion   c. autonomic system, afferent division   d. somatic system (afferent division)   e. somatic system, efferent portion

3. (LO 2) Specialized cells that play supportive roles in nervous tissue are called   a. glial cells   b. neurons   c. phagocytes   d. Nissl cells   e. fibrils

4. (LO 2) Ependymal cells   a. are phagocytes   b. form sheaths around CNS neurons   c. form sheaths about PNS cells   d. line the cavities of the brain and spinal cord   e. are sometimes called Schwann cells

5. (LO 3) Short, highly branched fibers specialized to receive impulses are called   a. axons   b. Nissl bodies   c. synaptic knobs   d. nodes of Ranvier   e. dendrites

6. (LO 3) In multiple sclerosis   a. patches of myelin deteriorate and are replaced by scar tissue   b. the patient may suffer from loss of coordination   c. the glial cells form neoplasms   d. answers (a) and (b) only are correct   e. answers (a),(b) and (c) are correct

7. (LO 4) In regeneration of an injured neuron the structure that guides the growth of new axon sprouts is the   a. cellular sheath   b. myelin sheath   c. Nissl substance   d. neurofilament   e. dendritic spine

8. (LO 5) Neurons that transmit impulses from the receptors to the CNS are   a. association neurons   b. motor neurons   c. sensory neurons   d. bipolar neurons   e. efferent neurons

9. (LO 6) A large bundle of axons within the spinal cord is a a. ganglion b. tract c. nerve d. nucleus e. endoneurium

10. (LO 7) You are driving along and a young girl darts in front of your car. The first step in the series of processes which must take place before you can hit the brake is a. integration b. conduction c. transmission d. reception e. effection

11. (LO 8) The patellar reflex a. is an example of a monosynaptic reflex b. requires only sensory and association neurons c. does not require an effector d. answers (a),(b) and (c) are correct e. answers (a) and (b) only are correct

12. (LO 8,9) Which of the following is NOT an essential component of a polysynaptic reflex pathway a. receptor b. afferent pathway c. brain d. efferent pathway e. effector

13. (LO 10) Which of the following is not considered a receptor a. muscle spindles b. olfactory epithelium c. structures that report pressure in the skin d. taste buds e. glands

14. (LO 11) The sodium pump a. equalizes the distribution of sodium in and outside of the neuron membrane b. actively transports sodium into the cell c. actively transports sodium out of the cell d. removes potassium from the cell e. answers (b) and and (d) are correct

15. (LO 12) A stimulus that increases the permeability of the neuron to sodium may result in a. resting potential b. propagation of an action potential c. polarization of sodium and potassium d. a refractory period that lasts up to 2 seconds e. breakdown of the all-or-none principle

16. (LO 13,14) A pain will be perceived as more intense if a. saltatory conduction is involved b. conduction is continuous rather than saltatory c. the impulse jumps along from one node to the next d. more neurons are stimulated e. two of the preceding are correct

17. (LO 15) John is suffereng from muscle spasm. His condition might be explained by a. lack of calcium b. lack of norepinephrine c. too much calcium d. the local procaine anesthetic his dentist has just injected e. DDT poisoning

18. (LO 16) Transmission of an impulse between two neurons usually depends upon a. release of transmitter substance b. chemicals released from dendritic spines c. saltatory conduction across the synapse d. continuous conduction of the action potential e. two of the preceding answers are correct

19. (LO 17) Norepinephrine is a. released in certain areas of the brain and spinal cord b. mainly inactivated by reuptake by vesicles in preganglionic synaptic knobs c. inactivated slowly by MAO d. released by motor neurons e. answers (a), (b) a and (c) are correct

20. (LO 18) Transmission is slowest in a. type A fibers b. unmyelinated fibers c. fibers of large diameter d. fibers that transmit impulses in two directions e. answers (a) and (c) are correct

21. (LO 19) Inhibitory postsynaptic potentials (IPSPs) a. cancel the effects of EPSPs b. hyperpolarize the postsynaptic membrane c. make the membrane more permeable to sodium d. answers (a) and (b) only are correct e. answers (a),(b) and (c) are correct

22. (LO 20,21) A neuron coming from the motor area of the brain synapses with hundreds of association neurons in the spinal cord. This is an example of a. convergence b. divergence c. afterdischarge d. a reverberating circuit e. facilitation

ANSWERS TO POST TEST QUESTIONS

1.c   2.b   3.a   4.d   5.e   6.d   7.a   8.c   9.b
10.d   11.a   12.c   13.e   14.c   15.b   16.d   17.a
18.a   19.e   20.b   21.d   22.b

9 THE NERVOUS SYSTEM
Divisions and Functions

PRETEST

1. From the outside in, the three meninges are the _____ _____, and _____ _____
2. The caudal tip of the spinal cord is the _____
3. In cross section the cord is seen to have a small central _____, surrounded by a butterfly shaped area of _____ _____
4. Two functions of the spinal cord are: (1) controls many _____ activities, and (2) transmits information back and forth to the _____ via its ascending and _____
5. The cerebrum and thalamus are derived from the embryonic _____
6. The brain and spinal cord are cushioned by the _____ fluid
7. The cardiac and vasomotor centers are located within the _____
8. The pneumotaxic center is located within the _____
9. Synaptic centers for visual and auditory neurons are found within the corpora quadrigemina of the _____
10. When sensory impulses reach the _____ we first begin to be aware of them
11. The hypothalamus is the link between nervous and _____ systems
12. The _____ acts to make muscular movements smooth instead of jerky
13. The second largest part of the brain is the _____
14. The most inferior portion of the brain stem is the _____
15. The cerebrum is divided into right and left halves by the _____ _____

16. The largest commissure, the _____
    links the neocortex of the two cerebral hemispheres
17. The motor cortex is located within the _____
    lobes
18. The visual cortex is located within the _____
    lobes
19. The caudate nucleus and lentiform nucleus are the
    two most prominent _____
20. About 80 percent of the fibers of the pyramidal
    pathway cross to the opposite side of the body within the pyramids of the _____
21. Voluntary movements are initiated by the _____
    areas of the _____ cortex
22. Pain from visceral structures is often _____
    to a superficial area
23. The limbic lobe is thought to be a link between
    cognitive and _____ mechanisms
24. When parts of the limbic system (hippocampal gyri)
    are removed a person can no longer _____
25. Environmental experience _____ cause
                                          can, cannot
    changes in brain structure
26. In most persons the _____ hemisphere is dominant for speech
27. When a person is studying, his brain waves are of
    the _____ type
28. The arousal system of the brain is the _____
29. Normal or slow wave sleep is known as _____
    sleep
30. The portion of the nervous system specialized to
    maintain internal homeostasis is the _____
    nervous system
31. Cranial nerve II is the _____ nerve
32. The dorsal root of a spinal nerve consists of
    _____ fibers while the ventral root consists
    of _____ fibers
33. The _____ portion of the autonomic system
    functions to mobilize energy during stress situations
34. The rate of the heart is speeded by _____
    nerves and slowed by _____ nerves
35. Drugs that have a stimulant effect because they
    block reuptake of norepinephrine and thereby enhance flow of impulses in the RAS are called _____

ANSWERS TO PRETEST QUESTIONS

1. dura mater, arachnoid, pia mater (LO 1) 2. conus medullaris (LO 2) 3. canal, gray matter (LO 3) 4. reflex, brain, descending tracts (LO 4) 5. forebrain (LO 5) 6. cerebrospinal (LO 6) 7. medulla (LO 7) 8. pons (LO 8) 9. midbrain (LO 9) 10. thalamus (LO 10) 11. endocrine (LO 11) 12. cerebellum (LO 12) 13. cerebellum (LO 13) 14. medulla (LO 13) 15. longitudinal fissure (LO 14) 16. corpus callosum (LO 14) 17. frontal (LO 15) 18. occipital (LO 15) 19. basal ganglia (LO 16) 20. medulla (LO 17) 21. motor, cerebral (LO 17) 22. referred (LO 18) 23. emotional (LO 19) 24. store new information (LO 19,20) 25. can (LO 21) 26. left (LO 22) 27. beta (LO 23) 28. RAS (reticular activating system) (LO 24) 29. non-REM (LO 25) 30. autonomic (LO 26) 31. optic (LO 27) 32. sensory, motor (LO 28) 33. sympathetic (LO 29) 34. sympathetic, parasympathetic (LO 30) 35. amphetamines (LO 31)

> LO 1: List the structures that protect the brain and spinal cord.

Brain and spinal cord are protected by bone, cerebrospinal fluid, and by the three meninges.

1. List the 3 meninges from the outside, in: (a)_____
   (b)_____
   (c)_____
2. What structures other than the meninges protect the brain_____

TEST YOURSELF

Match the following

1. dura mater____
2. pia mater____
3. arachnoid____
4. cerebrospinal fluid____

a. thin, vascular membrane that adheres to brain and cord
b. in subarachnoid space
c. lies just beneath subdural space
d. thick, tough, outer membrane

LO 2: Give the anatomic position of the spinal cord and describe its gross structure. (Label its structure upon a diagram.)

The spinal cord extends from medulla to the level of the second lumbar vertebra; it lies within the vertebral canal. The filum terminale and cauda equina extend below the conus medullaris.

Tell what each of the following is:
(a) conus medullaris_____
(b) filum terminale_____
(c) cauda equina_____
(d) dorsal fissure_____
(e) ventral median fissure_____

TEST YOURSELF

Match the following

5. cauda equina_____
6. filum terminale_____
7. conus medullaris_____
8. ventral median fissure_____

a. caudal tip of cord
b. extension of pia mater
c. spinal nerves that extend below the conus medullaris, within the vertebral canal
d. longitudinal groove
e. none of these choices is correct

LO 3: Describe and label on a diagram the structures of the spinal cord seen when the cord is viewed in cross section.

In cross section the central canal of the spinal cord may be seen. It is surrounded by gray matter and an outer portion of white matter. Each half of the white matter is arranged into anterior, posterior, and lateral funiculi.

1. Label the figure (on the following page) showing a cross section through the spinal cord.
2. In the figure indicate where each of the following would be found: (a) gray matter (b) white matter (c) spinal tracts (d) cell bodies of motor neurons

145

3. Logically, the dorsal fissure could also be called the posterior fissure. Similarly, the ventral fissure could be called the _____ fissure. The posterior funiculi could also be referred to as the dorsal funiculi. The anterior funiculi could be called the _____ funiculi.

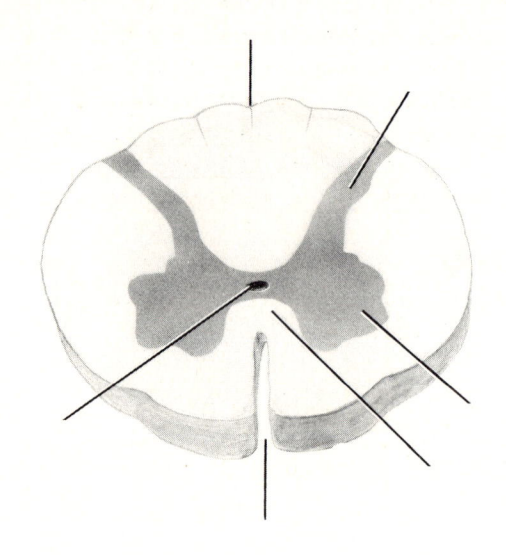

TEST YOURSELF
9. gray matter _____
10. funiculus _____
11. fasciculus _____

a. consists of large masses of cell bodies and dendrites
b. a tract
c. a column of the cord
d. a deep groove

LO 4: Give two functions of the spinal cord.

Two functions of the spinal cord are:
(1) _____
(2) _____

TEST YOURSELF
12. The spinal cord a. serves as a reflex control center b. transmits information to the brain c. transmits information from the brain to motor neurons d. answers (a), (b) and (c) are correct e. answers (b) and (c) only are correct

LO 5: Trace the development of the divisions of the brain from their origin in the neural tube of the embryo, and locate the ventricles.

In the early embryo the anterior end of the neural tube develops into forebrain, midbrain, and hindbrain. Forebrain later becomes telencephalon and diencephalon. The telenchephalon gives rise to the cerebrum, the diencephalon to the thalamus and hypothalamus. The midbrain gives rise to the corpora quadrigemina. The hindbrain differentiates into metencephalon which gives rise to the cerebellum and pons, and myelencephalon which becomes the medulla.

1. The lateral ventricles develop within the _____
2. The third ventricle develops within the _____
3. The cavity of the medulla is the _____

TEST YOURSELF
13. cerebrum ____   a. derived from myelencephalon
14. medulla ____   b. has lateral ventricles
15. cerebellum ____   c. derived from metencephalon

LO 6: Give the functions of the cerebrospinal fluid.

The CSF cushions the brain against mechanical shock and is though t to act as part of a brain barrier.

1. What are brain barriers _____
2. What is a choroid plexus _____

TEST YOURSELF

16. In which of the following might the CSF be important in protecting the brain   a. a child bumps his head   b. a man ingests arsenic   c. the choroid plexuses overproduce   d. answers (a),(b) and (c) are correct   e. answers (a) and (b) only are correct

LO 7-12: Give the function of the medulla, pons, midbrain, thalamus, hypothalamus, and cerebellum. Also give the general functions of the cerebrum.

Give the functions for each structure:

STRUCTURE	FUNCTIONS
Medulla	
Pons	
Midbrain	
Thalamus	
Hypothalamus	
Cerebellum	
Cerebrum	

TEST YOURSELF
17. medulla _____
18. pons _____
19. cerebrum _____
20. cerebellum _____
21. hypothalamus _____
22. midbrain _____

a. contains vital centers that control respiration, heart, rate, etc.
b. center of intellect, language
c. has pneumotaxic center; helps regulate respiration
d. regulates body temperature; thermostate
e. controls many visual and auditory reflexes
f. helps maintain posture and equilibrium

LO 13: On a given diagram label the divisions of the brain and the structures discussed.

Label the structures indicated (on the following page):

TEST YOURSELF

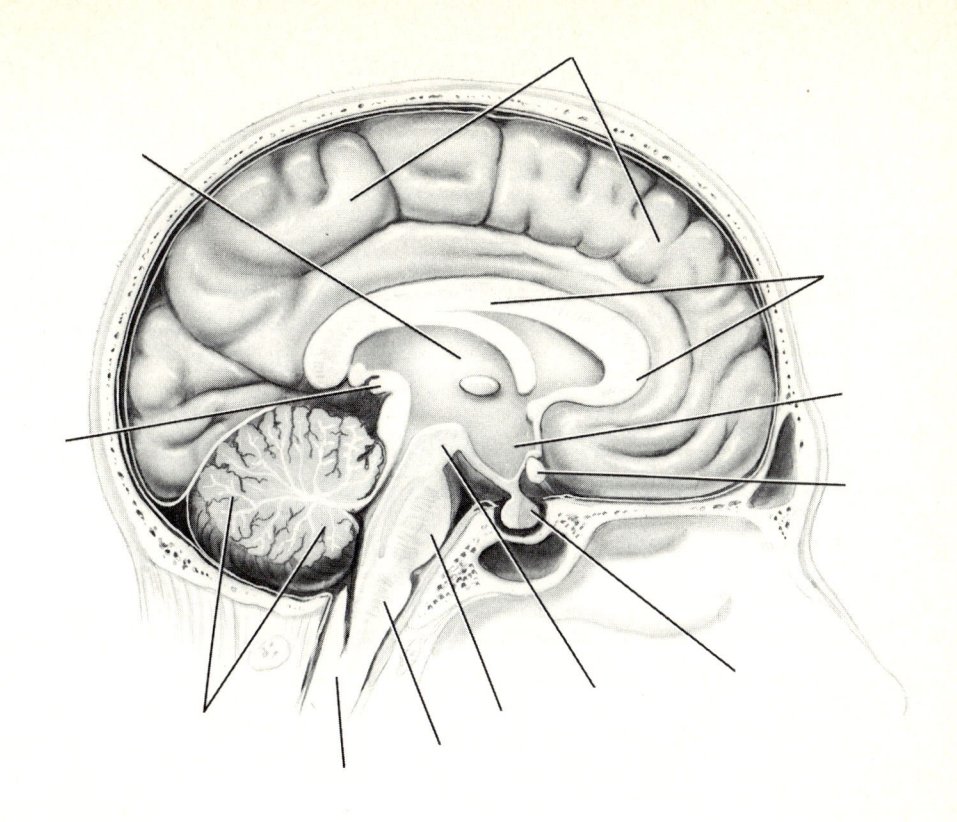

Match the following:
23. largest part of the brain___
24. second largest part of the brain___
25. most inferior part of brain (continuous with spinal cord___
26. part of brain stem just superior to medulla___

a. cerebellum
b. cerebrum
c. hypothalamus
d. medulla
e. pons
f. midbrain
g. thalamus

    LO 14: Describe the gross structure of the cerebrum.

The cerebrum is divided into right and left hemispheres by the longitudinal fissure. The cerebral cortex consists of groove-like gyri separated by groove-like gyri. Beneath the cortex is the white matter which contains the basal ganglia. Each hemisphere is divided into several lobes.

149

1. Define and indicate the location of e ch structure on the figure on p. 151.
   a) convolution_____
   b) gyrus_____
   c) sulcus_____
   d) fissure_____
2. On a separate sheet of paper draw a superior view of the cerebrum and label the longitudinal fissure.
3. Where is the white matter of the cerebrum located?
   _____

TEST YOURSELF
27. gyrus____   a. longitudinal fissure
28. cerebral cortex__   b. transverse fissure
29. sulcus____   c. gray matter
30. between cerebrum   d. deep groove
    and cerebellum__   e. shallow groove
                           f. convolution

LO 15: Identify the lobes of the cerebrum and their general functions.

1. Label the diagram on p. 151.
2. Give the functions for each lobe:

Lobe   Functions

Frontal_____

Parietal_____

Occipital_____

Temporal_____

Limbic_____

TEST YOURSELF
31. frontal   a. Broca's speech area
32. parietal   b. interpret visual stimuli
33. limbic   c. interpret auditory stimuli
34. temporal   d. concerned with emotional and sexual
35. occipital   behavior
                  e. receives and integrates information
                     from receptors in skin and joints

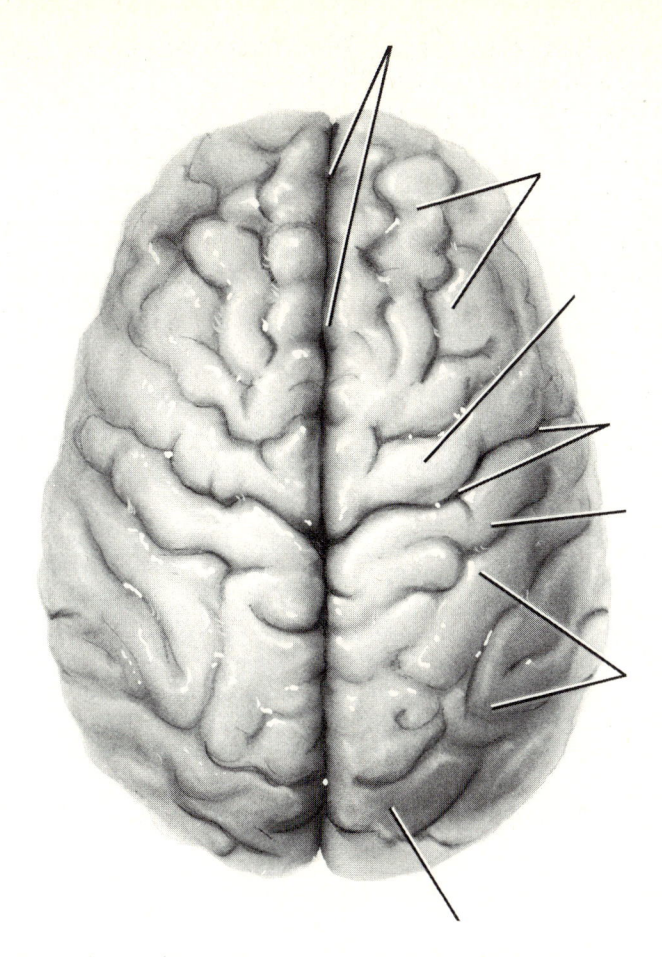

LO 16: Identify the basal ganglia and describe their function.

Basal ganglia are thought to be involved in planning and programming movement, but it is not known whether they actually generate movements or not.

1. Locate the basal ganglia_____
2. List two prominent basal ganglia_____

TEST YOURSELF

36. The basal ganglia a. are located within the white matter of the cerebrum b. are actually nuclei c. consist of gray matter d. answers (a),(b) and (c) are correct e. answers (a) and (c) only are correct

LO 17: Describe the control of movement and posture by the motor areas of the cerebral cortex, by other parts of the brain and cord, and by the nerves of the pyramidal and extra-pyramidal systems.

Specific areas of the precentral gyrus control voluntary movements of specific muscle groups. Messages regarding fine, skilled movements are sent via neurons of the pyramidal tracts to association neurons in the spinal cord. These synapse with motor neurons. Messages concerning posture or gross movements are sent via extrapyramidal tracts to the motor and association neurons in the cord.

1. What is dopamine_____
2. Describe paresis_____
3. Describe spastic paralysis_____

4. What is the role of the basal ganglia in control of movement_____

5. What is the role of the cerebellum_____
_____

TEST YOURSELF

37. Skilled learned movements are executed via a. the pyramidal pathway b. the extrapyramidal pathway c. the premotor area d. the prefrontal lobe e. the paresis

38. In Parkinson's disease a. there is a lack of dopamine in the caudate nuclei b. spastic paralysis is an important diagnostic symptom c. the cerebellum is damaged d. the pyramidal axons synapse directly with motor neurons e. none of the preceding is correct

    LO 18: Describe how we perceive sensation including pain.

Association areas are needed for the interpretation of sensory input received by primary sensory areas in the cerebral cortex.

1. Where does pain perception begin_____
2. What role does the limbic system play in the perception of sensation_____
3. What is referred pain_____
4. What causes phantom pain_____
   _____

TEST YOURSELF

39. If the primary visual cortices in both hemispheres are destroyed, one would a. see bright colors or flashes of light, but no objects b. see objects but no color c. be totally blind d. be totally deaf e. none of the preceding is correct

40. A person with a diseased heart feels intense pain in his left arm. He is experiencing a. phantom pain b. referred pain c. somatic pain d. loss of function in the spinothalamic tract e. malfunction of the thalamus

> LO 19: Identify the limbic system and its actions.

The limbic system includes the limbic lobe and several associated structures. It is a link between emotional and cognitive mechanisms, and is thought to affect emotional responses, sexual behavior, motivation, and storage of information.

1. What happens when electrodes are implanted in the "pleasure center" of the limbic system?_____
2. How is memory affected when the hippocampal gyri of the limbic lobes are removed_____
   _____

TEST YOURSELF

41. The limbic system is thought to function in a. memory storage b. emotional responses c. sexual behavior d. interpretation of sounds e. answers (a),(b) and (c) are correct

LO 20: Review current theories regarding learning and memory.

1. Briefly describe 3 theories of how memories are established (a)_____
   (b)_____
   (c)_____

2. What is the physiologic basis of retrograde amnesia
   _____
   _____

TEST YOURSELF

42. Memories are stored almost totally a. in the parietal lobe b. in the precentral gyrus c. in the prefrontal lobes d. in the cerebellum e. throughout the cerebral cortex, but not in a specific location

   LO 21: Cite experimental evidence linking environmental stimuli with demonstrable changes in the brain and with learning ability

Compare the brains of rats reared in an enriched environment with those of deprived rats
_____
_____
_____

LO 22: Tell how the two cerebral hemispheres differ in function.

1. In most persons the left hemisphere is best developed for_____

2. The right hemisphere appears better developed for
_____

TEST YOURSELF

44. When the left hemisphere is removed in young children a. aphasias occur b. the corpus callosum takes over its function c. the right hemisphere takes over its functions d. speech is no longer possible e. answers (a) and (d) are correct

> LO 23: Identify four main types of brain wave patterns.

Alpha waves are associated with relaxed states, beta waves with mental activity, delta with normal sleeping, and theta with emotional stress. In children theta waves are observed in normal situations.

1. What is an EEG _____
2. Can a person learn to control his brain waves _____
_____

TEST YOURSELF

45. Mary is resting with her eyes closed. What type of brain waves would she be producing   a. alpha   b. beta   c. delta   d. theta   e. excessive discharge

> LO 24: Describe the action of the RAS.

The RAS acts to regulate consciousness by feeding impulses into the cerebrum.

1. Give an example to demonstrate that the motor cortex sends messages to the RAS _____
_____

2. What happens when signals from the RAS become weak _____
_____

TEST YOURSELF

46. When the RAS is severely damaged, the victim   a. can no longer produce serotonin   b. is unable to sleep   c. goes into a deep, permanent coma   d. becomes paralyzed   e. beta waves become predominant

> LO 25: Summarize current views of sleep and discriminate between REM and non-REM sleep.

When the RAS transmits fewer signals the cerebrum gets sleepy. Sleep centers in the brain stem may also release serotonin which helps to bring on sleep.

1. Describe nonREM sleep in physiologic terms_____
2. Describe REM sleep_____
   _____

TEST YOURSELF

Match the following

47. REM sleep_____   a. normal sleep; metabolic rate
48. NonREM sleep_____   slowed
                          b. dreams occur

> LO 26: *Distinguish between somatic and autonomic nervous systems, and compare their efferent components.*

The somatic portion of the nervous sytem acts to preserve the integrity of the organism with reference to the external environment while the autonomic system helps maintain a steady state within the body. Somatic nerves innervate skeletal muscles; autonomic nerves innervate involuntary muscles and glands.

1. Which system--somatic or autonomic -- would act to
   a. inhibit digestion_____
   b. enable you to play a song on the piano
      _____
   c. speed the heart beat_____
   d. permit you to walk_____
   e. dilate the pupils_____

TEST YOURSELF

Match the following

49. innervation of sweat glands____   a. somatic system
50. pressure receptors in skin____
51. innervation of muscle in wall   b. autonomic system
    of stomch____
52. innervation of biceps muscles___

> LO 27: *List the twelve pairs of cranial nerves and give their functions.*

Cranial nerves transmit information to the brain from various receptors especially in the head region; their efferent components control movements of eyes, face, mouth, pharynx, larynx, heart, and other viscera.

A jingle such as the following may help you to remember the order of the cranial nerves. The first letter of each word is the first letter of the cranial nerve. (See Table 9-3)

<u>O</u>n <u>O</u>ld <u>O</u>lympus' <u>T</u>owering <u>T</u>ops <u>A</u> <u>F</u>inn <u>V</u>isited <u>G</u>reece <u>V</u>ending <u>S</u>ome <u>H</u>ats

1. In studying the cranial nerves you should list them in order and write the function for each, referring as necessary to Table 9-3

    For practice give the name and function for

	Name	Function
a. Cranial nerve I.	_____	_____
b. Cranial nerve II.	_____	_____
c. Cranial nerve VII.	_____	_____
d. Cranial nerve VIII.	_____	_____
e. Cranial nerve XII.	_____	_____

TEST YOURSELF

Match the following

53. Cranial nerve II ___   a. innervates heart
54. Cranial nerve V ___   b. Trigeminal
55. Cranial nerve IX ___   c. glossopharyngeal
56. Cranial nerve X ___   d. transmit messages from retina to brain

> LO 28: Describe (or label on a diagram) the structure of a spinal nerve.

Each spinal nerve has a ventral root and a dorsal root which is marked by a ganglion. Each nerve divides into dorsal and ventral rami and rami communicantes.

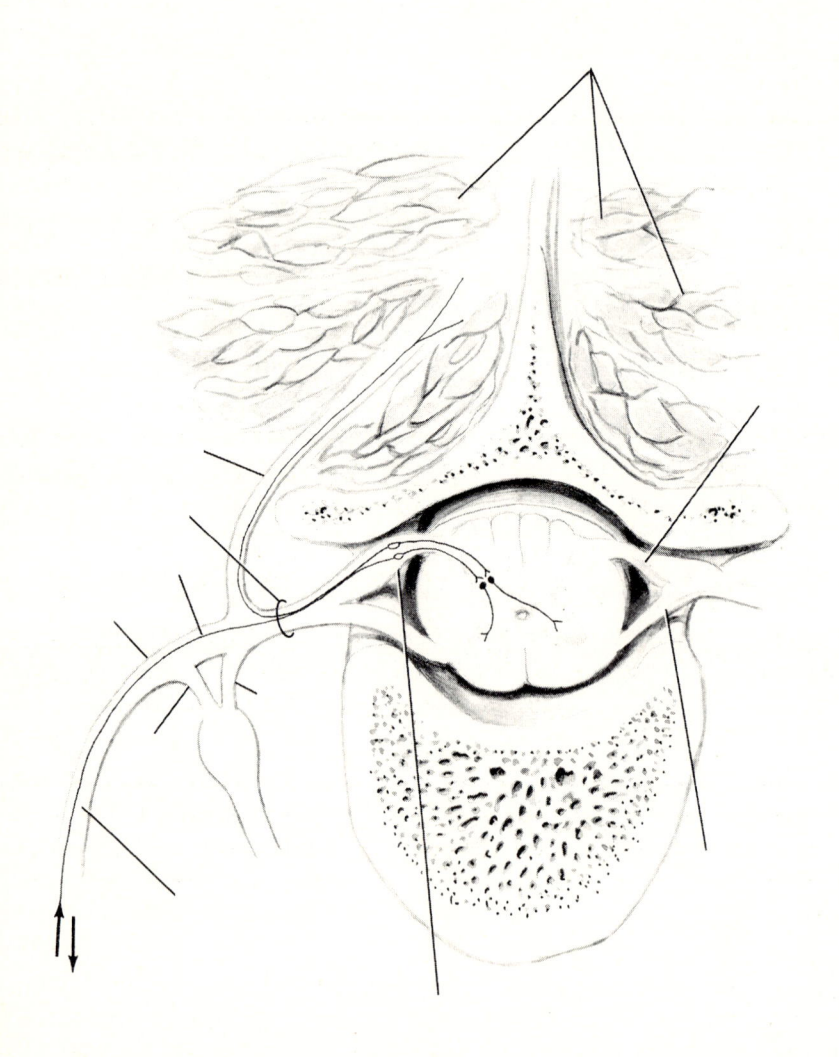

1. Label the diagram on preceding page
2. What type of nerve fibers run through the ventral root_____
3. The dorsal root_____
4. What is a ramus_____
5. What is a plexus_____

TEST YOURSELF

57. Nerve fibers that innervate muscles and skin of the posterior portion of the body would be found within
a. ventral rami   b. dorsal rami   c. meningeal rami
d. rami communicantes   e. ventral root

> LO 29: Contrast the sympathetic and parasympathetic systems.

Sympathetic nerves emerge at the thoracolumbar levels of the spinal cord, parasympathetic nerves at the craniosacral levels. The sympathetic system mobilizes energy during stress situations while the parasympathetic system acts to conserve and restore energy.

(See COMPARISON CHART on the following page)

List 4 ways in which the sympathetic system differs from the parasympathetic system. (For others consult chart on page 160 or table 9-6 in your textbook)

1._____
2._____
3._____
4._____

TEST YOURSELF

58. effect is widespread throughout the body_____   a. sympathetic system
59. nerves emerge at craniosacral level_____   b. parasympathetic system
60. chain and collateral ganglia _____

COMPARISON OF SYMPATHETIC WITH PARASYMPATHETIC SYSTEMS

	Sympathetic System	Parasympathetic System
Outflow from CNS	Thoracic, Upper Lumbar	Cranial (nerves III, VII, IX, and X) and Sacral
General effect	Mobilization of energy; lasting widespread effect	Conservation and restoration of energy; brief, localized effect
Preganglionic neuron	Extremely short	Long
Ganglia	Usually part of sympathetic chain located just lateral to vertebral column. Sometimes the ganglion is collateral (outlying). Sometimes the preganglionic neuron synapses in a ganglion above or below its level of outflow.	terminal ganglia (close to or within target organ)
Postganglionic neuron	Usually long; adrenergic	Very short; cholinergic
Some specific actions	Stimulates heartbeat	Inhibits heartbeat
	Dilates bronchial tubes	Constricts bronchial tubes
	Dilates pupil	Constricts pupil
	Inhibits intestinal motility	Stimulates intestinal motility
	Constricts cutaneous blood vessels	
	Stimulates sweat glands	

LO 30: List several organs innervated by both sympathetic and parasympathetic nerves and give the effect of each type of stimulation.

Many organs are innervated by both sympathetic and parasympathetic nerves. Generally, one system stimulates and the other inhibits a specific action.

Give the effect of each system upon the organs indicated:

Organ	Sympathetic action	Parasympathetic action
Heart	_____	_____
pupil of eye	_____	_____
intestine	_____	_____
bronchial tubes	_____	_____

TEST YOURSELF

61. dilates pupil
62. dilates bronchial tubes
63. inhibits intestinal motility

a. sympathetic system
b. parasympathetic system

LO 31: Identify types of drugs given in Table 9-8 and give their actions and effects upon mood.

Give the effect of each of the following drugs upon the nervous system:

Drug	Effect
Alcohol	_____
amphetamines	_____
barbiturates	_____
phenothiazines	_____

TEST YOURSELF

64. Which of the following are considered "major tranquilizers" and are used as antipsychotics
a. amphetamines   b. barbiturates   c. valium
d. morphine   e. phenothiazine

65. Which of the following inhibits impulse conduction in the RAS   a. amphetamines   b. caffeine   c. nicotine   d. barbiturates   e. answers (a),(b) and (c) are correct

ANSWERS TO TEST YOURSELF QUESTIONS

1.d	2.a	3.c	4.b	5.c	6.b	7.a	8.d	9.a
10.c	11.b	12.d	13.b	14.a	15.c	16.e	17.a	
18.c	19.b	20.f	21.d	22.e	23.b	24.a	25.d	
26.e	27.f	28.c	29.e	30.b	31.a	32.e	33.d	
34.c	35.b	36.d	37.a	38.a	39.c	40.b	41.e	
42.e	43.d	44.c	45.a	46.c	47.b	48.a	49.b	
50.a	51.b	52.a	53.d	54.b	55.c	56.a	57.b	
58.a	59.b	60.a	61.a	62.a	63.a	64.e	65.d	

POST TEST

1. LO 1) The thin, vascular membrane that adheres to the brain and spinal cord is the   a. dura mater   b. pia mater   c. arachnoid   d. periosteum   e. subarachnoid

2. (LO 2) The spinal cord   a. extends from the medulla at the level of the foramen magnum to the level of the second lumbar vertebra   b. has an average length of 48 inches   c. has a cephalic tip called the conus medullaris   d. gives rise to cauda equina at the cervical and lumbar enlargements   e. three of the preceding answers is correct

3. (LO 3) Ascending tracts are found within the   a. gray matter of the cord   b. filum terminale   c. funiculi   d. ventral median fissure   e. dorsal fissure

4. (LO 4) Which of the following are functions of the spinal cord a. receives sensory information from the skin receptors b. integrates incoming information from some sensory receptors c. dispatches messages to muscles d. sends information to the brain e. all of the preceding answers are correct

5. (LO 5) The fourth ventricle is found within the a. cerebrum b. cerebellum c. medulla d. pons e. hypothalamus

6. (LO 6) Cerebrospinal fluid is produced by the a. arachnoid villi b. gyri c. sulci d. transverse fissure e. choroid plexices

7. (LO 7-12) The part of the brain that helps maintain posture and equilibrium is the a. cerebrum b. pons c. basal ganglia d. thalamus e. cerebellum

8. (LO 7-12) The part of the brain that links nervous and endocrine systems is the a. thalamus b. cerebrum c. cerebellum d. hypothalamus e. pons

9. (LO 13) The portion of the brain located under the thalamus is the a. cerebrum b. pons c. hypothalamus d. cerebellum e. medulla

10. (LO 14) The corpus callosum is found within the a. thalamus b. medulla c. pons d. white matter of the cerebrum e. white matter of the hypothalamus

11. (LO 15) The visual association areas are found within which lobe a. frontal b. occipital c. temporal d. limbic e. central

12. (LO 16) The caudate nucleus is one of the a. basal ganglia b. pyramidal pathways c. lentiform nuclei d. gyri of the cerebrum e. largest commissures

13. (LO 17) Voluntary movements are initiated in the a. limbic lobe b. precentral gyrus c. hippocampal gyrus d. pons e. pyramids of the medulla

14. (LO18) In order to know that you are looking at a dog as opposed to a rock which structures must participate   a. primary visual cortex   b. visual association area   c. pyramidal tract   d. answers (a) and (b) are correct   e. answers (a),(b) and (c) are correct

15. (LO 19) A system that links emotional cognitive mechanisms and affects sexual behavior and information storage is the   a. pyramidal system   b. extrapyramidal system   c. reticular activating system   d. limbic system   e. prefrontal system

16. (LO 20) According to current theories memory is based upon   a. a change that takes place in the synaptic knobs or postsynaptic neurons   b. newly established circuits   c. the rhythm at which neurons fire   d. changes within glial cells   e. all of the preceding answers are correct

17. (LO 21) Physical changes can take place in the brain   a. only before birth   b. only during the first 6 months of life   c. in response to environmental experiences   d. only as dictated by the genes

18. (LO 22) Recognition of faces and objects based on shape is thought to be mainly the function of   a. the corpus callosum   b. right hemisphere   c. basal ganglia   d. pyramids   e. left hemisphere

19. (LO 23) You are sleeping peacefully (nonREM sleep). What type of brain waves would you emit   a. alpha   b. beta   c. delta   d. theta   e. none

20. (LO 24) The RAS causes activity in the   a. entire brain   b. medulla   c. pons   d. pyramids   e. answers (b) and (c) only

21. (LO 25) In REM sleep   a. breathing rate is slowed   b. metabolic rate is slowed   c. blood pressure is lowered   d. beta waves are emitted   e. all of the preceding answers are correct

22. (LO 26) In the somatic nervous system the number of neurons between CNS and effector is   a. one   b. two   c. three   d. the number varies   e. at least four.

23. (LO 27) The cranial nerve that innervates many of the viscera is   a. V   b. X   c. I   d. II   e. IX

24. (LO 28) The branches of spinal nerve are known as   a. plexi   b. rami   c. dorsal roots   d. commissures   e. foramina

25. (LO 29) Unlike the sympathetic system, the parasympathetic system has   a. a more localized effect   b. a more lasting effect   c. chain and collateral ganglia   d. norephinephrine as the transmitter substance released at synapse with the effector   e. all of the preceding answers are correct

26. (LO 30) Which of the following is(are) not stimulated by the sympathetic system   a. heart   b. intestine   c. adrenal medulla   d. sweat glands   e. answers (a), (c) and (d) only

27. (LO 31) Which drug has a stimulating effect because it enhances the flow of impulses in the RAS   a. alcohol   b. phenothiazine   c. amphetamines   d. barbiturates   e. cocaine

ANSWERS TO POST TEST QUESTIONS

1.b	2.a	3.c	4.e	5.c	6.e	7.e	8.d	9.c
10.d	11.b	12.a	13.b	14.d	15.d	16.e	17.c
18.b	19.c	20.a	21.d	22.a	23.b	24.b	25.a
26.b	27.c

10 THE SENSE ORGANS

PRETEST

1. A pacinian corpuscle is basically a _____ receptor
2. Sense organs in general function as _____, that is, devices that transfer and usually transform _____
3. The nerve ending of the pacinian corpuscle admits _____ ions in response to _____
4. The response of the pacinian corpuscle increases more than/ less than/ in proportion to the stimulus strength
5. Under unchanging or constant stimulation, the response of the pacinian corpuscle eventually _____
6. Free nerve endings respond to touch and pressure and seem to be able to produce the sensation of _____ and _____
7. Sensitive _____ _____ are touch receptors occuring in large numbers in the fingertips and lips.
8. The taste receptors of the anterior part of the tongue respond principally to _____ tasting substances
9. The cell bodies of the olfactory receptors are located in the _____ _____
10. The eyes are partly enclosed in the bony, conical _____ of the skull
11. The moist membranes covering the white of the eye and partly lining the anterior orbits are called the _____
12. The extrinsic muscles act to _____ the eyes, originating on the bones of the _____

166

13. What refracts most of the light entering the eye? _____
14. What fills the posterior chamber of the eye? _____
15. What is the light-sensing layer of the eye? _____
16. The lens originates from the _____ of the embryo, and continues to grow throughout life. This eventually results in the vision abnormality called _____
17. Color vision involves _____ different varieties of retinal _____ cells
18. In low-light vision, technically termed _____ vision, the _____ cells are active
19. The spot of clearest daytime vision, the _____ _____ contains few or no _____ cells
20. Both the _____ and the _____ contain cells sensitive to specific shapes and to motion in particular directions
21. In binocular depth perception, the slightly differing images from the two eyes appear to be compared in the _____ _____ of the brain
22. _____ rod and cone cells are connected to each optic nerve fiber
23. The most important mechanism whereby the eye adjusts to increased illumination is by mutual _____ on the part of the retinal cells. The _____, however, is involved in that it does regulate the amount of light entering the eye
24. Sound in air represents alternate _____ and _____ of the gas. Its pitch depends upon its _____, loudness upon the _____
25. The trumpet-shaped _____ of the ear gathers sound
26. The tiny middle earbones, or ossicles, consist of the _____, the _____ and the _____
27. Sound transducers are located in the _____ of the inner ear
28. The highest sound frequencies stimulate the organ of _____ nearest to the _____ window
29. The _____ transmits vibrations within the cochlea
30. The _____ and the _____ muscles regulate the transmittance of sound by the ossicle chain

167

31. Vibrations enter the cochlea via the_____ _____
32. Static equilibrium is maintained by the _____ and the_____, or more precisely, by the jelly-like_____they contain
33. _____flows in the semicircular canals in response to body motion, and this flow is detected by the_____ _____
34. The vestibular component of the_____nerve is connected to the_____nuclei
35. _____or sound localization is possible because of comparisons of the sound that is reported by both the two ears via nerve tracts that cross over within the brain to the_____cerebral hemispheres
36. The_____ _____ _____is responsible for a safety mechanism called the_____ stretch reflex which relaxes the associated muscle

ANSWERS TO PRETEST QUESTIONS

1. pressure (barely acceptable: touch) (LO 1) 2. transducers--energy (LO 1) 3. sodium--stretching (LO 1) 4. less than (LO 1) 5. declines (LO 1) 6. pain--itching (LO 2) 7. Meisnner's corpuscles (LO 2) 8. sweet (LO 3) 9. olfactory epithelium (LO 3) 10. orbits (LO 4) 11. conjunctiva (LO 4) 12. move--orbit (LO 4,5) 13. cornea (LO 5) 14. vitreous (LO 5) 15. retina (LO 5) 16. skin-presbyopia (LO 5) 17. three-cone (LO 6) 18. scotopic--rod (LO 6) 19. fovea centralis--rod (LO 6) 20. retina--visual cortex (LO 6) 21. geniculate nuclei (LO 7) 22. many (LO 7) 23. inhibition--iris (LO 5,7) 24. condensation--rarefaction--frequency--amplitude (LO 8) 25. pinna (LO 8) 26. ossicles--malleus--incus--stapes (LO 8) 27. cochlea (or scala media, or even membranous labyrinth) (LO 8,9) 28. corti--oval (LO 8,9) 29. endolymph (LO 9) 30. stapedius--tensor tympani (LO 9) 31. round window (LO 9) 32. saccule--utricle--macula (LO 9) 33. endolymph--crista macularis (cupula is acceptable) (LO 9) 34. auditory--vestibular (LO 10) 35. stereophony--contralateral (LO 11) 36. Golgi tendon organ--inverse (LO 11)

> LO 1: Describe the receptor potential, adaptation and response curve, using the pacinian corpuscle as an example of a typical sense organ.

The pacinian corpuscle consists of concentric layers wrapped around a nerve ending. Increased pressure produces increasing frequencies of nerve depolarization, but soon after stimulation has begun the nerve becomes less sensitive to it (adaptation), and in any case frequency of response is proportional only to mild stimulation, becoming less so at higher intensities. All these effects are produced by changes in the membrane potential of the sense organ itself (receptor potential).

1. What is a receptor potential? _____
   _____
2. At first, the receptor potential of a pacinian corpuscle is roughly _____ to the strength of the stimulus, but _____ with the passage of time
3. This last phenomenon is called _____
4. The response of the pacinian corpuscle to increased pressure is _____ than one might expect on the basis of its initial response
5. The actual response of a pacinian corpuscle to a pressure stimulus is _____ than one might expect, which has the effect of making a long-continued stimulus _____ to one's conscious mind
6. The responding portion of a pacinian corpuscle is the central _____ ending, which reacts in response to the stretching of its _____

7. What is the difference between a receptor potential and an action potential? _____
   _____

TEST YOURSELF

1. Adaptation involves   a. a decline of response frequency at high stimulation strength   b. a decline of response frequency with the passage of time   c. the inability to continue to perceive a long-continued stimulus of unvarying strength   a. (a) and (b)   e. (b) and (c)

> LO 2: Summarize the diffuse sense organs by name, location and function. (For summary see table 10-1)

169

Give the function, in each case:
1. free nerve endings_____
2. Merkel's discs_____
3. Ruffini's end organs_____
4. Meissner's corpuscles_____

TEST YOURSELF

2. detect joint angulation
3. pressure sensitive; stimulated by very rapid tissue movement
4. pain receptors
5. sensitive fingertip touch receptors

a. pacinian corpuscle
b. Ruffini's end organ
c. Meissner's corpuscle
d. Golgi tendon organ
e. free nerve ending

LO 3: Describe the receptors of smell and taste.

The taste buds are sensitive to the primary stimuli of sweet, bitter, sour and salty. The smell receptors are located in the olfactory epithelium of the roof of the nasal cavity. Taste buds are composed mostly of non-nervous cells which continuously regenerate; olfactory receptors are neurons whose cell bodies occur in the olfactory epithelium itself.

1. The_____ _____ are the basic receptors of the sense of taste. They are mostly located in the_____
2. Taste buds differ in their taste sensitivity; for example, "sweet receptors" occur mainly at the _____ of the tongue
3. The olfactory nerve fibers penetrate the_____ of the_____bone
4. The olfactory receptors are especially vulnerable to injury because the cell bodies are located in the _____
5. Describe olfactory adaptation_____
_____

170

TEST YOURSELF

6. The taste bud is able to respond to _____ basic taste sensation, and the olfactory receptors to perhaps _____   a. 4-50   b. 50-4   c. 4-2   d. 25-25

7. Taste buds are   a. nerve cell bodies which, when destroyed, are lost permanently   b. non-nervous structures   c. capable of regeneration   d. lost in old age   e. (b),(c) and (d)

> LO 4: Describe the location, anatomic relations, and structure of the eyes.
>
> LO 5: Give the function and describe the normal action of the following:
> a. extrinsic muscles
> b. cornea
> c. iris
> d. lens (with emphasis upon accomodation and image formation)
> e. aqueous
> f. vitreous
> g. retina

Light enters the eyes through the cornea, a transparent window in the sclera. The entrance of light is regulated by the iris, focused by the lens, and sensed as an image by the retina. The eye is kept turgid by the presence of the aqueous fluid and vitreous body, both transparent to light. The elastic lens is focussed by contraction of the ciliary muscle but the light entering the eye is mostly refracted by the cornea.

1. Be sure to review the special section "light and the eye". When you have done so, answer the following:
   a. What mostly determines the extent to which light is refracted by a transparent substance? _____

   b. What structure mostly refracts light which enters the eye? _____
   c. Why is the retinal image of an object inverted? _____

d. What is the shape of the lens in near vision? _____ far? _____

2. Label this illustration

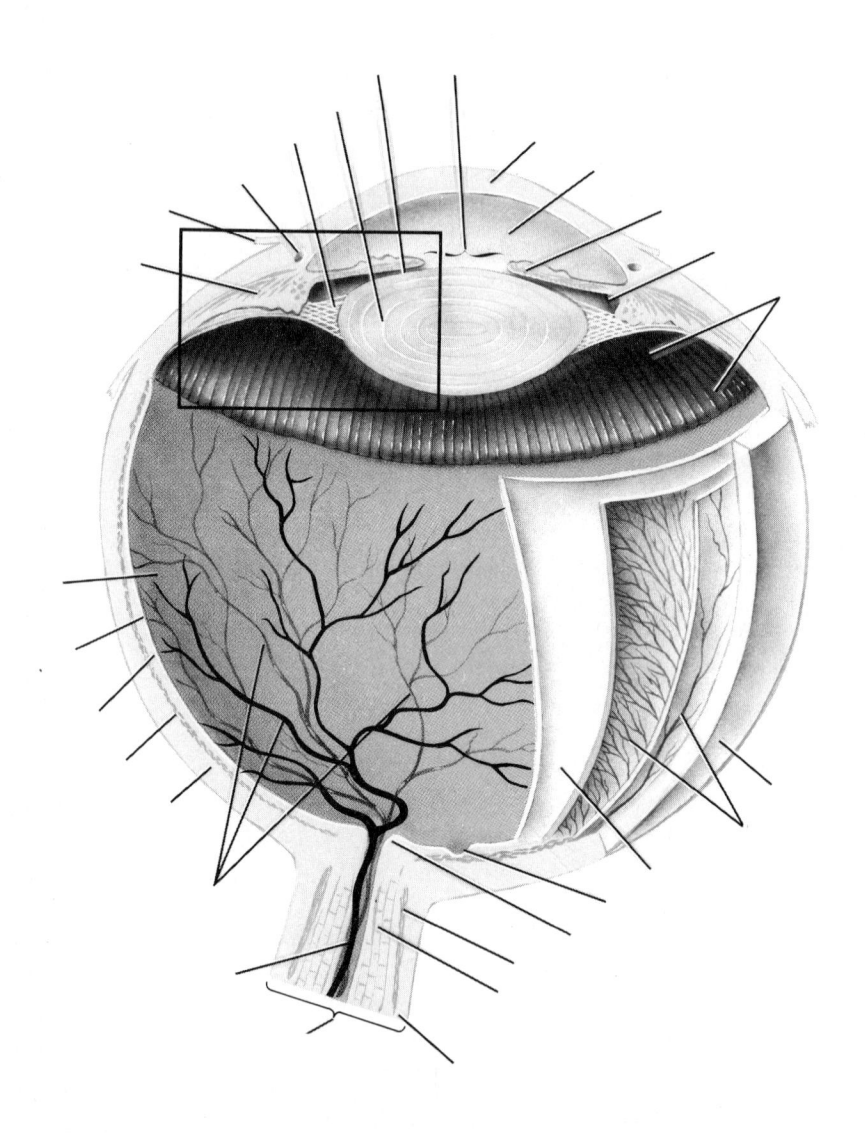

172

3. Consulting figs. 10-4 and 10-5 list agonists and antagonists involved when the RIGHT eye:

Action	agonists	antagonists
Looks to right		
Looks to left		
Looks upward		
Looks downward		
Looks left and down		

4. (a) What produces the aqueous?_____
   (b) Where does it normally go?_____
   (c) What would you say is its normal function?_____
   (d) What happens in glaucoma?_____

5. (a) The_____are little fibers acting as the insertion of the_____muscle upon_____of the lens
   (b) When the ciliary muscle relaxes, the lens becomes_____; when its fibers contract, the lens becomes_____
   (c) Give the technical terms for:
      nearsightedness_____
      farsightedness_____
      senile inability to accomodate_____
   (d) What produces presbyopia?_____

6. The iris is_____by adrenergic drugs, and_____by cholinergic ones. Its function in part, is to_____light entering the eye. Its behavior also reflects _____states

173

TEST YOURSELF

8. admits light to the eye
9. dilates in low light and strong emotion
10. produces most of the refraction of light in the eye
11. focusses light
12. fills anterior chamber
13. light-sensitive layer

a. cornea
b. lens
c. iris
d. aqueous
e. retina

14. Glaucoma could be produced by a. excessive production of aqueous b. blockage of the canal of Schlemm c. dilation of the retina d. all of the preceding e. (a) and (b) only

15. Which of these structures is located INSIDE the eyeball? a. lacrimal duct b. canal of Schlemm c. conjunctiva d. medial rectus muscle e. levator palpebrae muscle

>LO 6: Summarize photopic and scotopic vision, relating each to the sensory cells of the retina and their characteristics.

Photopic vision is based upon the action of cone cells. It functions at high light intensities and enables one to see color. Scotopic vision is a property of the dark-adapted rods. They do not perceive color. There seem to be three kinds of cones, each maximally sensitive to a different wavelength of light (corresponding roughly to a primary color). The depolarization of both rod and cone cells results from the light-dependent breakdown of rhodopsin-like visual pigments.

1. Photopic vision is the ability to see at _____ light intensities and scotopic is the ability to see at _____ levels of illumination
2. Signify superiority with a + sign, inferiority with a - sign:

	photopic	scotopic
sensitivity		

	photopic	scotopic
color vision		
acuity		

3. Label this diagram:

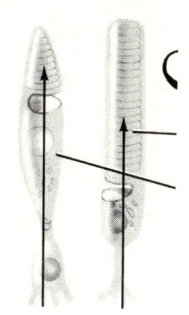

4. Why does vitamin A deficiency produce night blindness?

TEST YOURSELF

16. The cone cells   a. are active in scotopic vision b. are commonest near the periphery of the visual field and almost entirely absent from the foves centralis c. occur in three frequency-sensitive varieties d. periodically shed their intracellular discs e. all of the preceding

> LO 7: Describe two roles of the retina and of the brain in control and interpretation of vision.

The interpretation of vision begins in the retina through the mutual inhibition and other interaction of light sensitive cells which, for example, minimizes the effect of light scattering within the retina and also adjusts retinal sensitivity to prevailing levels of illumination. Moreover, specific simple interpretation of shapes, lines and edges appear to take place in the retina itself. The retinal output of information is centrally processed in the brain in a variety of ways involving a variety of neuronal arrangements. These have further specific sensitivities to edges, lines, directions, etc. with final interpretation occuring in the visual cortex and the cerebellum (which coordinates visual information with proprioception amd motor function). Depth perception largely results from a comparison by the geniculate nuclei of the slightly differing images seen by the two *eyes though these differences are normally suppressed and fused into a single picture in our visual consciousness. Apparently by statistical calculation and comparisons, the brain perceives color and improves the quality of visual information reported to it by the retina. The iris, ciliary muscle and extrinsic muscles are all more or less under reflex or unconscious control of the brain.*

1. What kinds of preliminary visual interpretaion are known (or believed) to take place in the retina itself?_____
   _____

2. The_____receives visual input which it employs in muscular coordination, especially of walking

3. The_____ _____, composed in man of alternating layers of neurons, compares the slightly_____images from the two eyes, resulting in_____perception. On the conscious level, however, inhibition fuses the two into a _____ _____

4. The ciliary muscles are controlled by the_____ reflex, and binocular_____upon some object is also, in part, a reflex

5. The conscious perception of vision occurs in the _____ _____of the brain (see also chapter 9)

TEST YOURSELF

17. Integration of visual and proprioceptive information for motor coordination occurs in the   a. geniculate nuclei   b. corpus callosum   c. optic chiasma   d. infundibulum   e. cerebellum

18. The eyes response to varying levels of illumination is regulated by the   a. iris   b. ciliary muscle   c. retina   d. geniculate nucleus   e. (a) and (c)

LO 8: Describe the structure of the ear.

The external ear consists of a pinna and a canal (meatus) closed on the blind end by the eardrum (tympanic membrane). The middle ear contains the malleus, incus and stapes, tiny bones connected to the drum and to the inner ear. The middle ear is also connected to the pharynx by the eustachian tube. The inner ear consists of cochlea and vestibular apparatus. The cochlea, or organ of hearing, consists of a bony portion, the scalae tympani and vestibuli, and a membranous portion called the scala media.

Label the following diagram:

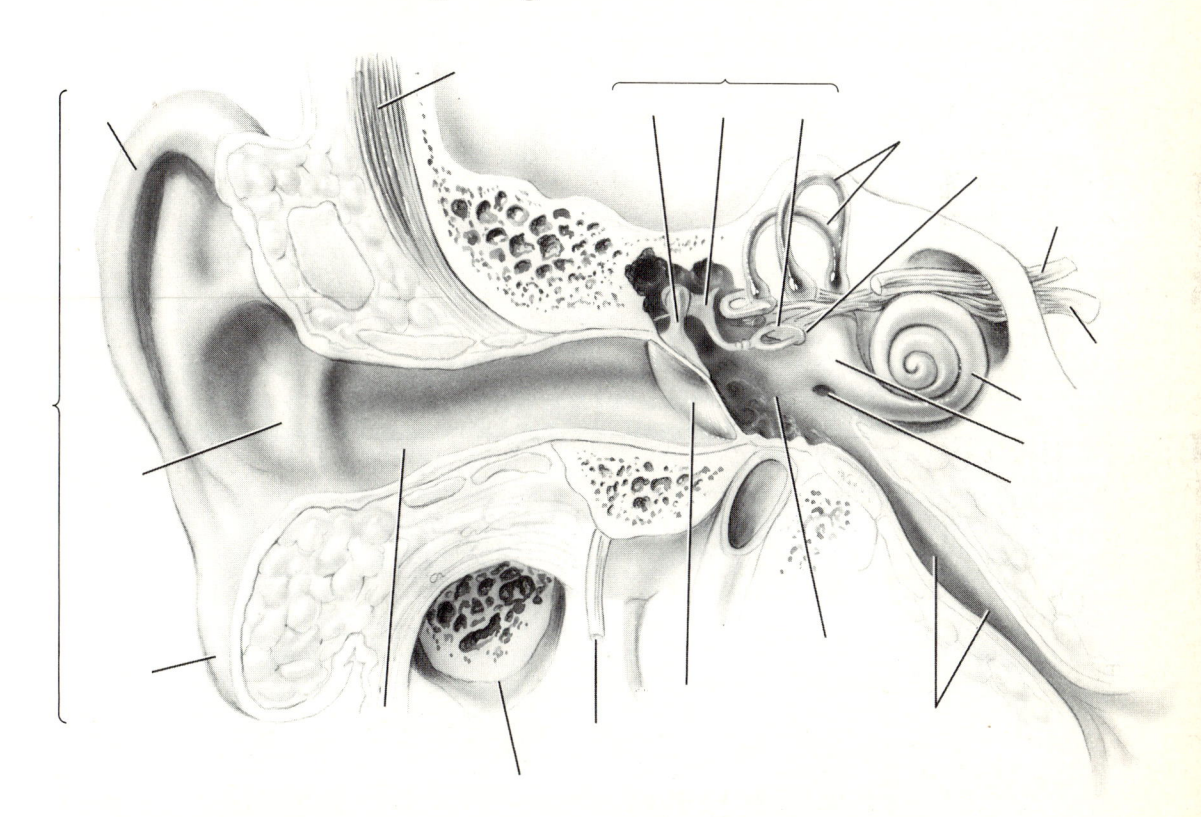

TEST YOURSELF
19. external trumpet of the ear
20. structure that first vibrates in sympathy of sound waves
21. cerumen secreted here
22. chain of tiny middle ear bones
23. snail-shaped structure of inner ear

a. meatus
b. cochlea
c. drum (tympanic membrane)
d. ossicles
e. pinna

24. entrance to cochlea
25. exit from cochlea
26. contains otoconia
27. contains cupulae
28. contains organ of Cori

a. vestibular apparatus
b. ampulla
c. scala media
d. oval window
e. round window

    LO 9: Give the function and describe the
    normal action of the following, in relation
    to the physical nature of sound:
    a. external ear and ear canal
    b. middle ear
        (1) ossicles
        (2) eustachian tube
        (3) drum
    c. inner ear
        (1) cochlea
        (2) vestibular apparatus

Atmospheric vibrations are transmitted from the eardrum (or tympanic membrane) after first having been gathered by the pinna and been channeled by the meatus. They proceed along the chain of ossicles to the oval window of the cochlea. There they are transferred to the scalae and exit via the round window. The organ of Corti is located within the scala media; its hair cells respond to sound vibrations. The organ of Corti responds regionally to differential frequencies, with the lowest being sensed at the apex near the helicotrema. Distortions of the hair cells produce receptor potentials that travel via the auditory nerves to the lower auditory centers and auditory cortex of the brain.

Static equilibrium is sensed by the macula. Its otoconia are displaced by motion of the head in relation basically to the pull of gravity. This displacement is transmitted in the form of distoriton of the jelly in which both they and the hair cells are embedded. The hair cells respond with receptor potentials that are transmitted via the vestibular component of the auditory nerve. The semicircular canals contain endolymph which responds to changes in the direction of motion by flowing over the cristae ampullaris which respond via their hair cells.

1. Label the following diagram:

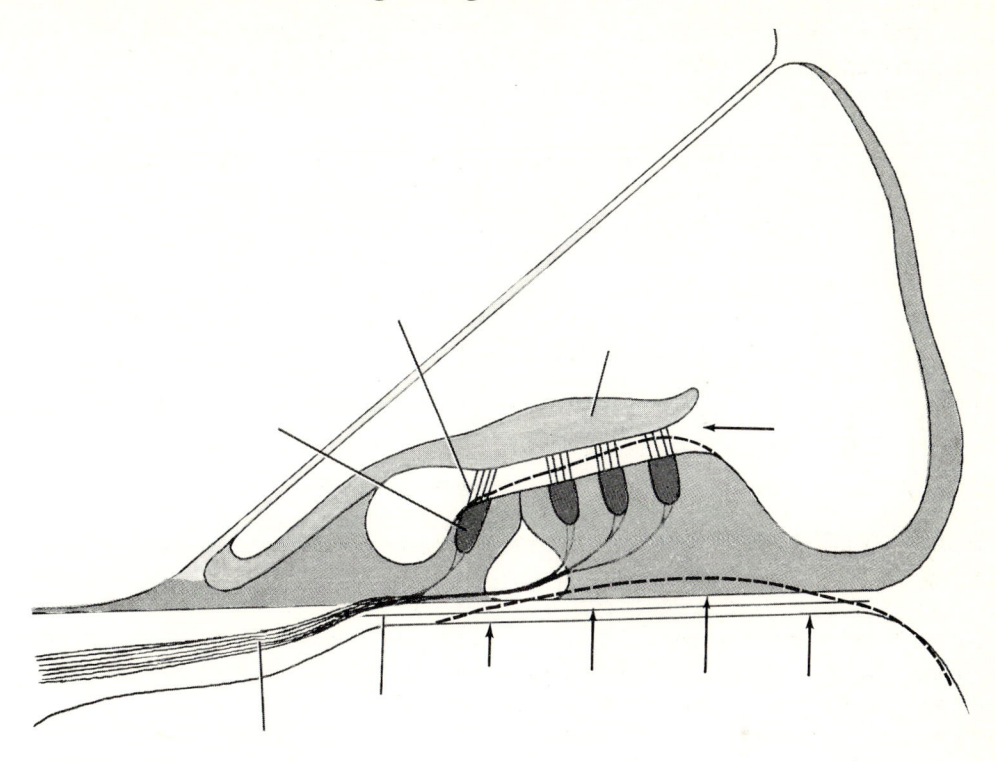

2. The term <u>frequency</u>, as applied to sound, refers to the _____ of the atmospheric vibrations
3. The term <u>amplitude</u>, on the other hand, refers to _____
4. In addition to the mere transmission of sound, the ossicles also _____
5. How does the cochlea detect differences in pitch? _____
6. The eustachian, or auditory tube acts to _____ pressure on the two sides of the eardrum
7. What is the function of the two muscles which are inserted on the malleus and stapes? _____

8. Describe the action of the tectorial membrane
   _____
   _____

9. Assign each of the following a number to indicate the order in which it transmits or otherwise responds to sound: (a) oval window   (b) stapes
   (c) round window   (d) malleus   (e) scala media
   (f) tympanic membrane

10. (a) The vestibular apparatus responds to gravity by movement of the stony _____
    (b) This movement distorts the jelly-like _____
    _____
    (c) That distortion is sensed by tiny_____

11. There are_____semicircular canals. The _____in these responds to changes in motion by flowing   past the_____ _____which is thereby distorted. This distortion is again sensed by_____ _____

TEST YOURSELF

29. articulates with oval window
30. regulates sound transmission by ossicles
31. equalizes pressure
32. rubs on hair cells in the membranous labyrinth
33. senses dynamic equilibrium and motion

a. stapedius
b. stapes
c. organ of Corti
d. crista ampullaris
e. eustachian tube

34. Which of the following sense organs does NOT employ hair cells?
    a. vestibular apparatus
    b. crista ampullaris
    c. membranous labyrinth
    d. olfactory epithelium

   LO 11: Summarize the malfunctions of vision and hearing given in the textbook.

Nearsightedness or myopia is generally caused by an eyeball which is abnormally long. Its opposite, farsightedness or hyperopia, results from a short eyeball. In both cases the image forms inappropriately. Presbyopia is a defect of accomodation arising in middle age from continued growth of the lens. Glaucoma is a condition of increased intraocular pressure produced by excessive aqueous production or the blockage of its absorption. Cataract is an opacity of the lens or the cornea.

Deafness may result from a mechanical defect of the transmission system of the ear; nerve deafness from a defect of the cochlea, auditory nerve or the interpretive structures of the brain. Presbycusis is a progressive developing insensitivity to high frequency sounds.

1. Describe:
   a. myopia_____
   b. hyperopia_____
   c. presbyopia_____
   d. glaucoma_____
   e. cataract_____
2. a. What is conduction deafness?_____
   b. What structures might be involved in conduction deafness?_____
   c. What is presbycusis and how is it thought to be caused?_____
3. Review the development of glaucoma (taken up in the textbook in connection with the production and circulation of aqueous).

TEST YOURSELF

36. myopia
37. presbyopia
38. presbycusis
39. glaucoma (closed angle)
40. conduction deafness

a. progressive loss of sensitivity to high frequency sounds
b. retinal blockage of the trabeculae
c. enlarged lens
d. fusion of ossicles
e. long eyeball

LO 12: Outline the role of the muscle spindle and Golgi tendon organ in the reflex control of muscular contraction.

Muscle effort is controlled by feedback originating in the muscle spindle sense organs. The Golgi tendon organ appears to be principally a safety device protecting the tendon from breakage by means of the inverse stretch reflex that inhibits muscular contraction.

1. When a muscle contracts, its muscle spindle fibers _____. This tends to _____ the spindle fiber, if the muscle as a whole is not shortening
2. The above is then sensed by the _____
3. Such stretching will be most marked in isotonic/isometric muscle contraction
4. In response, the force of the muscle contraction will be increased/decreased
5. On a separate sheet of paper explain the familiar knee jerk reflex in terms of the action of muscle spindle fibers
6. What is a Golgi tendon organ? _____

TEST YOURSELF

41. The golgi tendon organ a. responds strongly when isometric contraction occurs, producing a reflex increase in strength b. produces a protective inverse stretch reflex c. contracts along with the muscle of which it is a part d. controls the stapedius and tensor tympani muscles

ANSWERS TO ALL TEST YOURSELF QUESTIONS

1.e   2.b   3.a   4.e   5.c   6.a   7.e   8.a   9.c
10.a 11.b 12.d 13.e 14.d 15.b 16.c 17.e
18.e 19.e 20.c 21.a 22.d 23.b 24.d 25.e
26.a 27.b 28.c 29.b 30.a 31.e 32.c 33.d
34.d   (NOTE: an alternative version of this question might be: Which of the following does not respond to mechanical forces?) 35.b   36.e   37.c   38.a   39.b
40.d   41.b

POST TEST

1. (LO 1) As pressure increases, the pacinian corpuscle produces a receptor potential which
a. also increases in direct proportion to the pressure   b. increases, but more slowly than the pressure   c. stays the same in amplitude but increases in frequency   d. decreases steadily over a long period of time   e. increases both in amplitude and in frequency

2. (LO 1) Suppose now that the pressure stimulation is great enough initially to produce a response, and remains the same for a time. During that period which of the answers to No. 1 would apply?

3. (LO 2: Which of the following is a specialized pain receptor?   a. free nerve endings   b. Merkel's discs   Ruffini's end organs   d. Meissner's corpuscles   e. Golgi tendon organ

4. (LO 5) Which of these is particularly noted for regeneration?   a. taste buds   b. hair cells (of inner ear)   c. olfactory cells   d. brain neurons   e. retinal rod cells

5. (LO 5) Which of the following structures is/are transparent?   a. retina   b. trabecula   c. sclera   d. cornea   e. iris

6. (LO 5) Near accomodation, in which the eye foccuses upon some close object, involves   a. contraction of the ciliary muscle   b. relaxation of zonule tension   c. rounding up of the lens   d. all of the preceding

7. (LO 5,11) Glaucoma involves, or may involve
a. excess intraocular pressure   b. overproduction of aqueous   c. defective absorption of aqueous   d. blockage of the trabaculae by the iris   e. all of the preceding

8. (LO 6) Scotopic vision a. is least acute at the fovea b. involves receptor cells of three distinctive spectral sensitivities c. involves cells that shed their discs d. is highly sensitive to vitamin A deficiency e. THREE of the preceding

9. (LO 7) Which of the following is essential for stereoscopic depth perception a. nerve fiber decussation b. geniculate nuclei c. comparison of images d. coordinated action of intrinsic eye muscles e. (a),(b) and (c) but not (d) f. (a) and (c) only

10. (LO 8) Which of the following is a middle ear structure a. pinna b. ceruminous glands c. tensor tympani d. scala media e. round window

(Nos. 11-15 relate to LO 9)
11. dynamic equilibrium
12. respond to direction of gravity
13. transmit vibrations to round window
14. rubs against hair cells
15. amplify vibrations by leverage

a. otoconia
b. organ of Corti
c. ossicles
d. crista ampullaris
e. none of these

16. (LO 10) Which of the following process impulses transmitted by the auditory nerve a. vestibular nucleus b. auditory cortex c. cerebellum d. all of the preceding e. (a) and (b) only

17. (LO 11) Which of the following disorders results from the continued normal growth of a body part a. emmetropia b. presbyopia c. myopia d. presbycusis e. glaucoma

18. (LO 11) Which of the following disorders results from diabetes mellitis, among other things (review question pertinent io this chapter) a. presbycusis b. inability to taste c. inability to smell d. cataract e. none of the preceding

19. (LO 12) Which of the following is responsible for the knee jerk reflex (produced by striking the tendon of the quadriceps) a. muscle spindle b. Golgi

tendon organ   c. cerebellum   d. pacinian corpuscle
e. organ of Corti

ANSWERS TO POST TEST QUESTIONS

1.b   (NOTE: Be sure to distinguish between receptor potential and the response of the associated nerve fiber. This rules out choices c and e. Similar comments apply also to the next question.)   2.d   3.a
4.a   5.d   6.d   7.e   8.b   9.e   10.c   11.d
12.a   13.e   (NOTE: one of the choices resembles a correct answer very closely. But check it out carefully and you will see why it is incorrect)
14.b   15.c   16.d   17.b   18.d   19.a

11 PROCESSING FOOD
The Gastrointestinal System

PRETEST

1. On its way to the stomach, food passes through the long, muscular _____
2. When small enough, nutrient molecules can pass through the wall of the GI tract and into the blood This is the process of _____
3. The lining of the GI tract is called the _____
4. The _____ is a double fold of peritoneal tissue that attaches the small intestine to the posterior abdominal wall
5. In the mouth salivary amylase begins the digestion of _____
6. The roof of the mouth is the _____
7. Each tooth is composed mainly of a calcified connective tissue called _____
8. When allowed to accumulate bacteria that inhabit the mouth form _____ _____ upon the surfaces of the teeth
9. The rhythmic waves of contraction that move food along through the GI tract are called _____
10. The exit of the stomach is the _____
11. Hydrochloric acid is released by large _____ cells
12. Digestion of _____ begins in the stomach
13. An open sore in the wall of the stomach is known as a(n) _____ _____
14. From the stomach chyme passes into the _____
15. The fingerlike projections in the intestinal lining are _____
16. Three enzymes produced by in the small intestine are _____ _____ _____

186

17. The function of trypsin and chymotrypsin is _____
18. The liver secretes _____
19. _____ cells remove bacteria and old blood cells that pass through the liver sinusoids
20. Bile pigments are breakdown products of _____
21. Cholecystokinin (CCK) is released from the intestine mainly when _____ is present in the duodenum
22. The major product of carbohydrate digestion is _____
23. Lipase acts on _____ digesting them to _____ and _____
24. The goal of protein digestion is to hydrolyze _____ bonds releasing free _____
25. Within each villus is a network of _____ and a central lymph vessel called a _____
26. Glucose and amino acids are absorbed by _____
27. From the ascending colon wastes pass into the _____
28. Between the taenia the wall of the large intestine bulges outward as pouches called _____
29. Three functions of the large intestine are _____

ANSWERS: 1.esophagus (LO 1) 2.absorption (LO 2) 3.mucosa (LO 3) 4. mesentery (LO 4) 5.polysaccharides (carbohydrates) (LO 5) 6.palate (LO 6) 7.dentin(LO 7) 8.dental plaque (LO 8) 9.peristalsis (LO 9) 10.pylorus (LO 10) 11.parietal (LO 11) 12.proteins (LO 12) 13.peptic ulcer (LO 13) 14.duodenum (LO 14) 15.villi (LO 15) 16.peptidase,lipase, disaccharidases (maltase, sucrase, or lactase) (LO 16) 17.protein digestion (LO 17) 18. bile (LO 18) 19.Kupffer (LO 19) 20.hemoglobin (LO 20) 21.fat (LO 21) 22.glucose (LO 22) 23.triglycerides, glycerol, fatty acids (LO 23) 24.peptide, amino acids (LO) 25. capillaries, lacteal (LO 25) 26.active transport (LO 26) 27.transverse colon (LO 27) 28.haustra (LO 28) 29. absorption of water and sodium, incubation of bacteria, elimination of wastes (LO 29)

LO 1: List in sequence each structure through which a bite of food passes as it makes its way through the GI tract. (Be able to label the structures on a diagram.)

1. List the structures in sequence:

   mouth ⟶

2. Label the diagram on page 189.

TEST YOURSELF

Imagine that food is passing through the structure listed in the question column. Select the structure through which it would pass next from the answer column.

1. duodenum____
2. pharynx____
3. transverse colon___

a. esophagus
b. jejunum
c. ileum
d. ascending colon
e. descending colon

LO 2: Describe the general terms the following steps in processing food: ingestion, digestion, absorption, elimination.

Briefly describe
(1) Ingestion_____

(2) Digestion_____

(3) Absorption_____

(4) Elimination_____

TEST YOURSELF
Match the following
4. ingestion
5. digestion
6. absorption
7. elimination

a. fate of undigested food
b. each step requires a specific enzyme
c. chewing and swallowing
d. passage of nutrients into blood

189

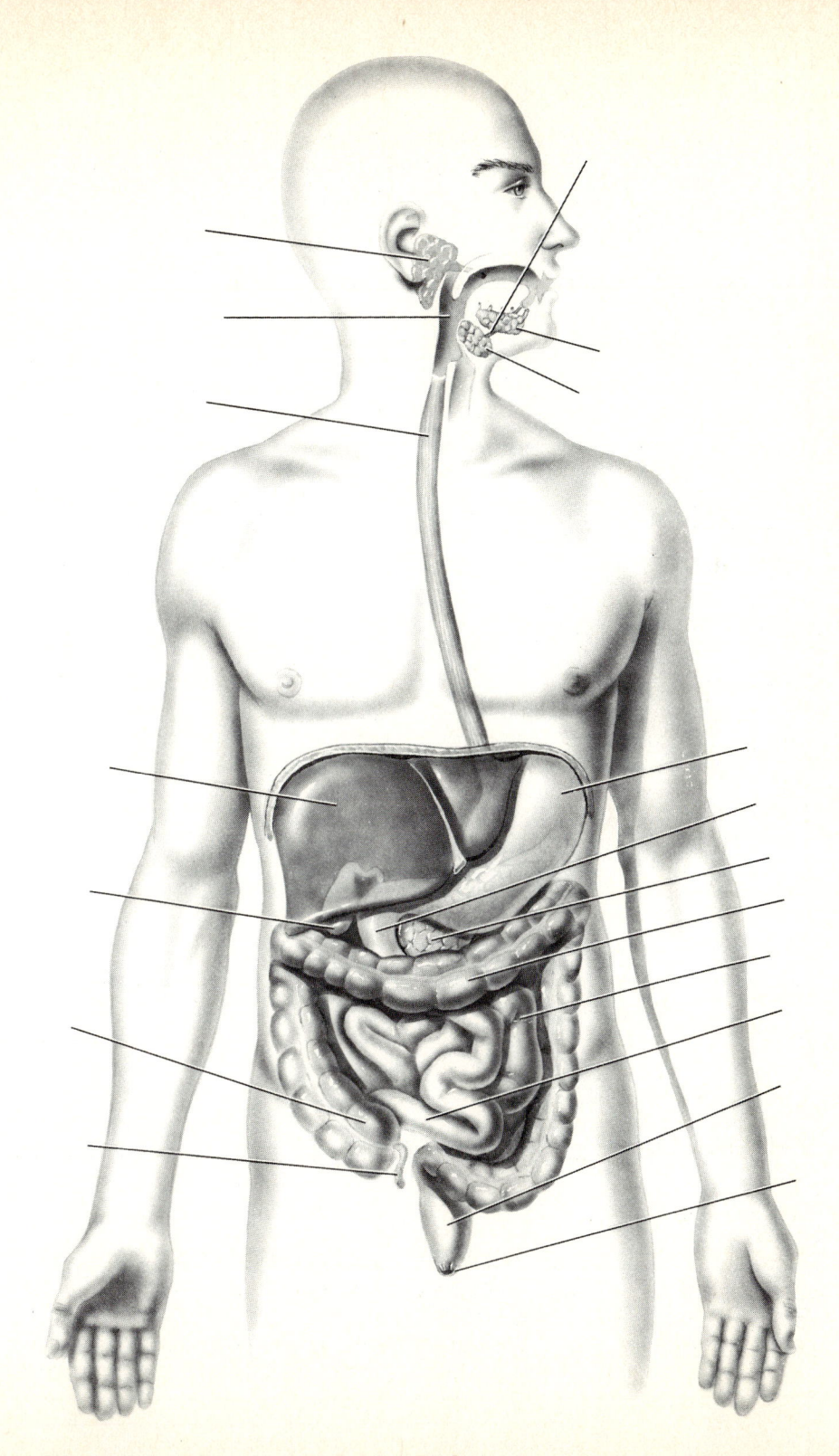

> LO 3: Identify and describe the four layers of the wall of the GI tract.

The layers are mucosa, submucosa, muscularis, and outer connective tissue covering.

Briefly describe each layer:
(1) Mucosa_____

(2) Submucosa_____

(3) Muscularis_____

(4) Outer layer_____

(5) What is the myenteric plexus?_____

(6) What is the submucosal plexus?_____

TEST YOURSELF

Match the following

8. epithelium lying upon connective tissue and a thin layer of muscle
9. circular and longitudinal sheets of muscle

a. muscularis
b. mucosa
c. submucosa

> LO 4: Distinguish between the visceral and parietal peritonea and describe their major folds.

Below the diaphragm the outer connective tissue covering of the GI tract is the visceral peritoneum. It connects by folds to the parietal peritoneum.

Describe:
(1) parietal peritoneum_____

(2) mesentery_____

(3) greater omentum_____

(4) lesser omentum_____

TEST YOURSELF

Match the following

10. sheet of connective tissue which lines abdominal wall
11. hangs down over the intestine like apron

a. parietal peritoneum
b. mesentery
c. greater omentum
d. lesser omentum

> LO 5: Describe the changes that take place in a bite of food as it is processed in the mouth.

In the mouth food is mechanically digested, moistened, tasted, and fashioned into a bolus. Polysaccharides are partially digested by salivary amylase to maltose and dextrin.

(1) Mechanical digestion in the mouth is accomplished by_____
(2) Tasting is the function of_____
(3) Food is moistened by the_____
(4) What is a bolus?_____

TEST YOURSELF

12. Which of the following occurs in the mouth
    a. mechanical digestion   b. chemical digestion
    c. taste   d. moistening   e. all of the preceding answers are correct

> LO 6: Describe the structures of the mouth and give their functions. Include the functions of saliva.

In the mouth food is chewed, moistened, tasted, and fashioned into a bolus. The chemical digestion of carbohydrates begins as salivary amylase acts on starch in the food.

(1) The roof of the mouth is the_____
(2) The fungiform and circumvallate papillae are found on the_____ and bear_____

(3) List 4 functions of saliva: 1._____
    2._____ 3._____ 4._____

TEST YOURSELF

13. Which of the following is not a function of saliva?
    a. lubricates mouth b. moistens food c. initiates carbohydrate digestion d. initiates protein digestion e. washes away food particles and bacteria

> LO 7: Draw and label a tooth and give the functions of each part.

1. On a separate sheet of paper draw a tooth and label its parts. (Consult Fig. 11-5 in your textbook.)
2. Give the function of:
   (a) pulp cavity_____
   (b) dentin_____
   (c) enamel_____
   (d) cementum_____
   (e) odontoblasts_____

TEST YOURSELF

Match the following
14. produce dentin   a. pulp
15. hardest substance in body   b. enamel
16. part of tooth that projects   c. odontoblasts
    into the socket   d. crown
                                      e. root

> LO 8: Describe the etiology of the most common disorders of the mouth: dental caries and periodontal disease.

Describe the development of
(a) dental caries_____

(b) periodontal disease_____

TEST YOURSELF

Match the following

17. dental caries _____   a. gums recede and bleed
18. peridontal disease _____ b. fluorides make teeth
                                 resistant

> LO 9: Describe the process of swallowing by
> tracing a bite of food to the stomach:
> Describe peristalsis.

In swallowing, reflex movements propel the bolus through the throat and into the esophagus. Peristaltic contractions push the food through the esophagus to the stomach.

(1) What mechanisms act to prevent entrance of food into the respiratory system during swallowing?
_____

(2) What type of tissue lines the esophagus? Why?
_____

TEST YOURSELF

19. During swallowing the opening to the respiratory tube is closed by the   a. tonsil   b. adenoid   c. epiglottis   d. fauces   e. uvula

> LO 10: Draw and label a diagram of the
> stomach.

On a separate sheet of paper draw the stomach and label its parts.

TEST YOURSELF

Match the following
20. main portion of stomach _____   a. fundus
21. exit of stomach _____   b. pyloric sphincter
22. ring of muscle _____   c. pylorus
                                      d. antrum
                                      e. corpus

> LO 11: Describe the microscopic structure
> of the stomach mucosa.

The simple columnar epithelial cells that line the stomach secrete mucus. Each of the millions of gastric glands contains specialized chief cells that secrete pepsinogen, large parietal cells that secrete hydrochloric acid, and neck cells which secrete mucus.

(1) How does the layer of oblique muscle fibers in the muscularis enable the stomach to better perform its job? _____

(2) What do the tiny pits in the stomach mucosa lead to? _____

(3) What are the granules seen in chief cells? _____
_____
_____

TEST YOURSELF

23. The stomach is lined with   a. stratified squamous epithelium   b. stratified columnar epithelium   c. oblique muscle   d. simple columnar epithelium   e. parietal cells

> LO 12: Summarize the functions of the stomach. (Trace the changes that occur in food as it passes through the stomach.)

(1) List 5 functions of the stomach
   1.
   2.
   3.
   4.
   5.

(2) Describe changes that occur in the food while it is in the stomach.

TEST YOURSELF

24. Which of the following is NOT a function of the stomach?   a. protein digestion   b. lipid digestion into free fatty acids   c. mechanical digestion of food   d. release of intrinsic factor   e. absorption of water

LO 13: Describe the protective mechanisms that prevent the gastric juice from digesting the stomach wall, and explain what happens when they fail (i.e. development of ulcers.)

Normally, the gastric juice does not digest the stomach wall because the thick coating of alkaline mucus secreted by the epithelial cells of its mucosa protect the tissues. Also the close arrangement of the epithelial cells prevent gastric juice from leaking between them and reaching underlying tissues. A third mechanism is the rapid turnover of epithelial cells lining the wall. Should these cells be damaged they are quickly replaced.

(1) What happens when these mechanisms fail?_____

(2) What effect does alcohol or aspirin have?_____

(3) Peptic ulcers most commonly occur in the_____

(4) Duodenal ulcers appear to be related to_____
_____

TEST YOURSELF

25. The thick coating of alkaline mucus that covers the gastric mucosa   a. helps prevent digestion of the stomach wall by the gastric juice   b. contains enzymes for digestion of cholesterol   c. helps in vitamin D absorption   d. answers (a), (b) and (c) are correct   e. answers (a) and (b) only are correct

LO 14: Trace the passage of chyme into the duodenum and describe the changes that occur in the chyme as it passes through the small intestine.

Small amounts of chyme are ejected into the duodenum from the stomach. In the duodenum bile from the liver and pancreatic juice mix with the chyme and help to digest food molecules.

(1) Most chemical digestion takes place in the_____

(2) As nutrients are digested they are_____
    in the blood
(3) Food that is not absorbed moves through the
    _____and_____and then into the
    large intestine

TEST YOURSELF

26. Food that is not absorbed moves from the duodenum
    into the   a. ileum   b. transverse colon   c. cecum
    d. jejunum   e. villi

> LO 15: Describe the anatomic features in the
> small intestine that increase its surface
> area, and describe its glands.

1. Three structures that increase the surface area of
   the small intestine are
   (1)_____
   (2)_____
   (3)_____
2. The intestinal glands are located between the
   _____. They secrete large amounts of
   _____.
3. Brunners glands secrete a thick_____

TEST YOURSELF

27. The surface area of the small intestine is increased
    by   a. villi   b. microvilli   c. plicae circulares
    d. answers (a),(b) and (c) are correct   e. none of
    the preceding answers is correct

> LO 16: List (in general terms) the functions
> of the small intestine.

Four functions of the small intestine are:
1._____

2._____

3._____

4._____

TEST YOURSELF

28. Which of the following is NOT a function of the small intestine? a. produces hormones b. absorption of nutrients c. produces bile d. receives pancreatic juice from the pancreas e. produces enzymes

> LO 17: Give the functions of the digestive enzymes secreted by the pancreas.

Give the function of each enzyme:

Enzyme	Function
Trypsin (also chymotrypsin and carboxypeptidase)	_____
Pancreatic lipase	_____
Pancreatic amylase	_____
Cholesterol esterase	_____
RNAase	_____
DNAase	_____

TEST YOURSELF

29. digests protein____
30. digests carbohydrate____
31. cholesterol____

a. lipase
b. cholesterol esterase
c. trypsin
d. pancreatic amylase
e. RNA-ase

> LO 18: List several functions of the liver that are associated with nutrition.

List 5 functions of the liver:

1. _____

2. _____

3. _____

4. _____

5. _____

TEST YOURSELF

32. Which of the following is NOT a function of the liver?   a. secretes bile   b. converts glucose to glycogen   c. stores iron   d. inventories nutrients   e. secretes enzymes

> LO 19: Describe the gross and microscopic structure of the liver.

The liver is divided into right and left lobes. Each lobe is divided into thousands of lobules consisting of plates of liver cells that radiate out from a central vein. Between the liver cells the bile canaliculi receive bile and empty it into larger bile ducts.

(1) Label the diagram on the following page (198)

(2) Trace the flow of blood through the liver. Blood is brought to the liver from the intestine via the hepatic portal vein. It then flows into the _____ _____. Blood is also brought to the liver by the hepatic arteries. The hepatic sinusoids empty into the _____ _____. These in turn empty into the larger_____ the_____ _____ that carry blood back toward the heart.

(3) What is the function of the Kupffer cells?

(4) Trace the flow of bile through the liver: liver cells to bile_____ then to right or left hepatic duct. The cystic duct from the_____ joins the common hepatic duct to form the common_____ which opens into the _____

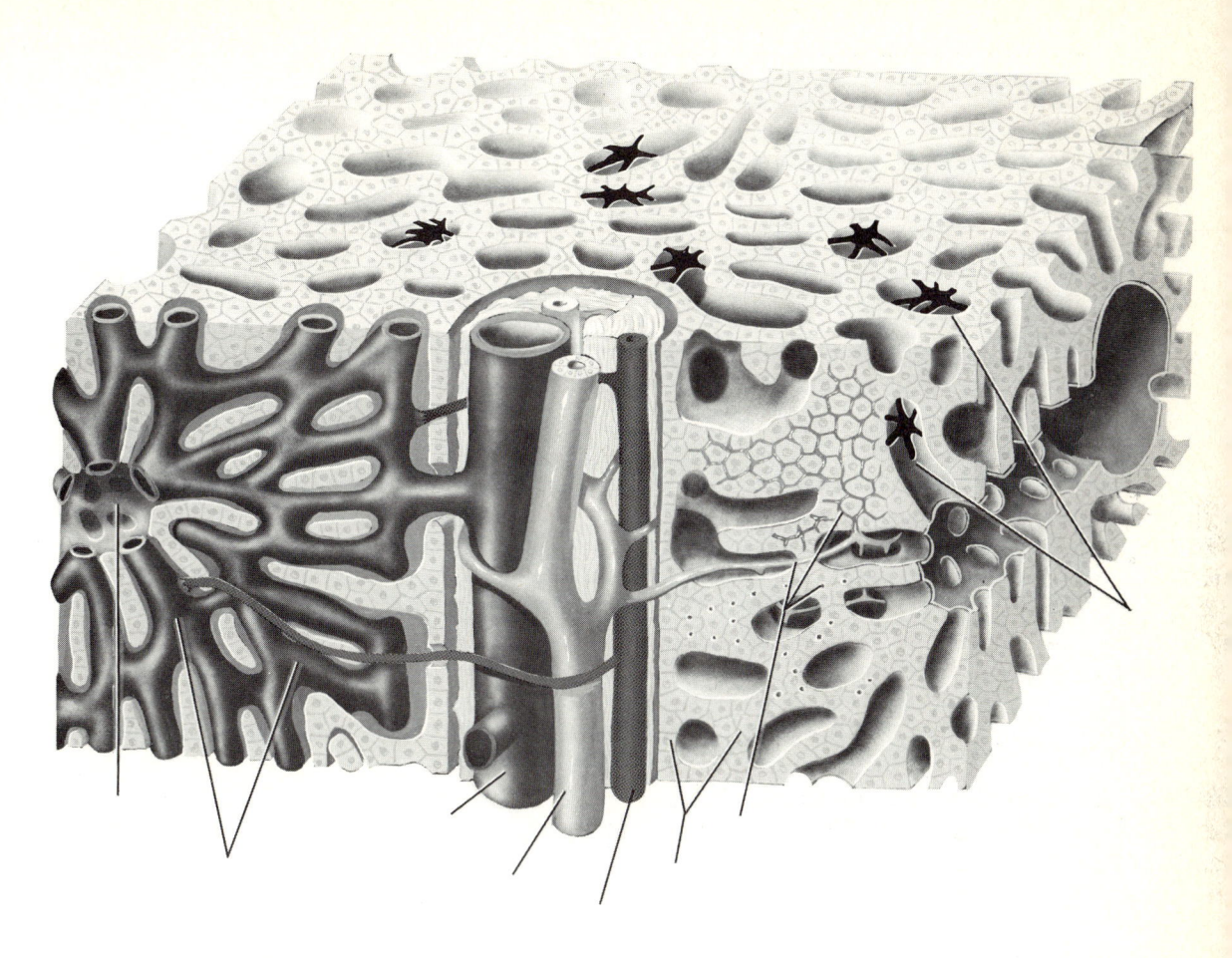

TEST YOURSELF

33. Blood flows into the liver sinusoids from the
    a. hepatic portal vein b. hepatic arteries c. central veins d. bile canaliculi e. answers (a) and (b) are correct

34. The tiny ducts that convey bile from the liver cells to the larger bile ducts that run between the lobules are called a. common bile ducts b. sinusoids c. bile canaliculi d. cystic ducts e. right or left hepatic ducts

> LO 20: Give the composition of bile, and describe its action.

199

Bile is composed of water (97%), bile salts, bile pigments, cholesterol, electrolytes, and lecithin.

(1) Give the fuction of the bile salts_____
(2) What is bilirubin?_____
(3) Where do the bile pigments come from?_____

(4) Why is cholesterol present in the bile?_____

(5) What happens when cholesterol precipitates in the liver ducts or gallbladder?_____

(6) What causes jaundice?_____
(7) What are micelles?_____

35. Which of the following are functions of the bile ? a. excretion of bile pigments b. emulsification of fats c. digestion of carbohydrates d. answers (a), (b) and (c) are correct e. answers (a) and (b) only are correct

> LO 21: Give the function of the gallbladder and describe its regulation by cholecystokinin. (CCK)

When the sphincter of Oddi is closed bile is shunted into the gallbladder for storage. When cholecystokinin is secreted it stimulates the gallbladder to contract releasing bile into the ducts leading to the duodenum.

(1) Under what conditions is CCK released?_____

(2) Where is CCK produced?_____
(3) What happens to bile in the gallbladder?_____

TEST YOURSELF

36. Cholecystokinin (CCK) is a. a hormone b. released by the liver c. produced by the intestinal mucosa d. released in response to protein in the intestine e. Two of the preceding answers is correct

LO 22: Summarize carbohydrate digestion in a step by step fashion indicating the location of each reaction and the name of the enzyme as well as the breakdown product.

In carbohydrate digestion polysaccharides are broken down into monosaccharides, simple sugars that are small enough to be easily absorbed.

Describe in words or equations what happens to a carbohydrate in:
(1) the mouth _____
(2) the stomach _____
(3) duodenum (a) action of pancreatic juice _____

(b) action of enzymes in brush border _____

TEST YOURSELF

37. maltose hydrolyzed to glucose ____
38. amylase inactivated ____

a. in stomach
b. by amylase
c. by enzyme from brush border of duodenum
d. in pancreas

LO 23: Summarize the step by step digestion of lipids.

In lipid digestion fats are degraded into free fatty acids and glycerol.

Describe what happens to lipids in
(1) the mouth _____
(2) stomach _____
(3) duodenum (a) action of bile _____

(b) action of pancreatic juice _____

TEST YOURSELF

39. Chemical hydrolysis of triglycerides depends mainly upon the action of   a. pepsin   b. bile   c. pancreatic lipase   d. salivary lipase   e. cholesterol esterase

LO 24: Summarize the step by step chemical
digestion of proteins.

In protein digestion proteins are hydrolyzed so that free amino acids are released.

Describe the fate of proteins in:
(1) the mouth_____
(2) stomach_____
(3) duodenum (a) action of pancreatic juice____
_____
            (b) action of enzymes in brush border__
_____

TEST YOURSELF
Match the following
40. protein degraded to polypeptides in   a. stomach
41. small polypeptides and dipeptides   b. pancreas
    hydrolyzed to free amino acids in   c. brush border
                                               d. mouth

LO 25: Describe the structure of an intestinal villus and draw a diagram of a villus labeling its parts.

An intestinal villus is a fingerlike projection of the mucus membrane which contains a capillary network and a lacteal. Nutrients are absorbed through its single layer of epithelial cells and then enter the capillary or lacteal.

Label the diagram on the following page (p.203)

TEST YOURSELF

42. Which of the following are part of the structure of an intestinal villus a. stratified squamous epithelium b. a network of capillaries c. fat tissue d. mucus-secreting glands e. all of the preceding answers are correct

LO 26: Describe the absorption of glucose, amino acids, and fat.

203

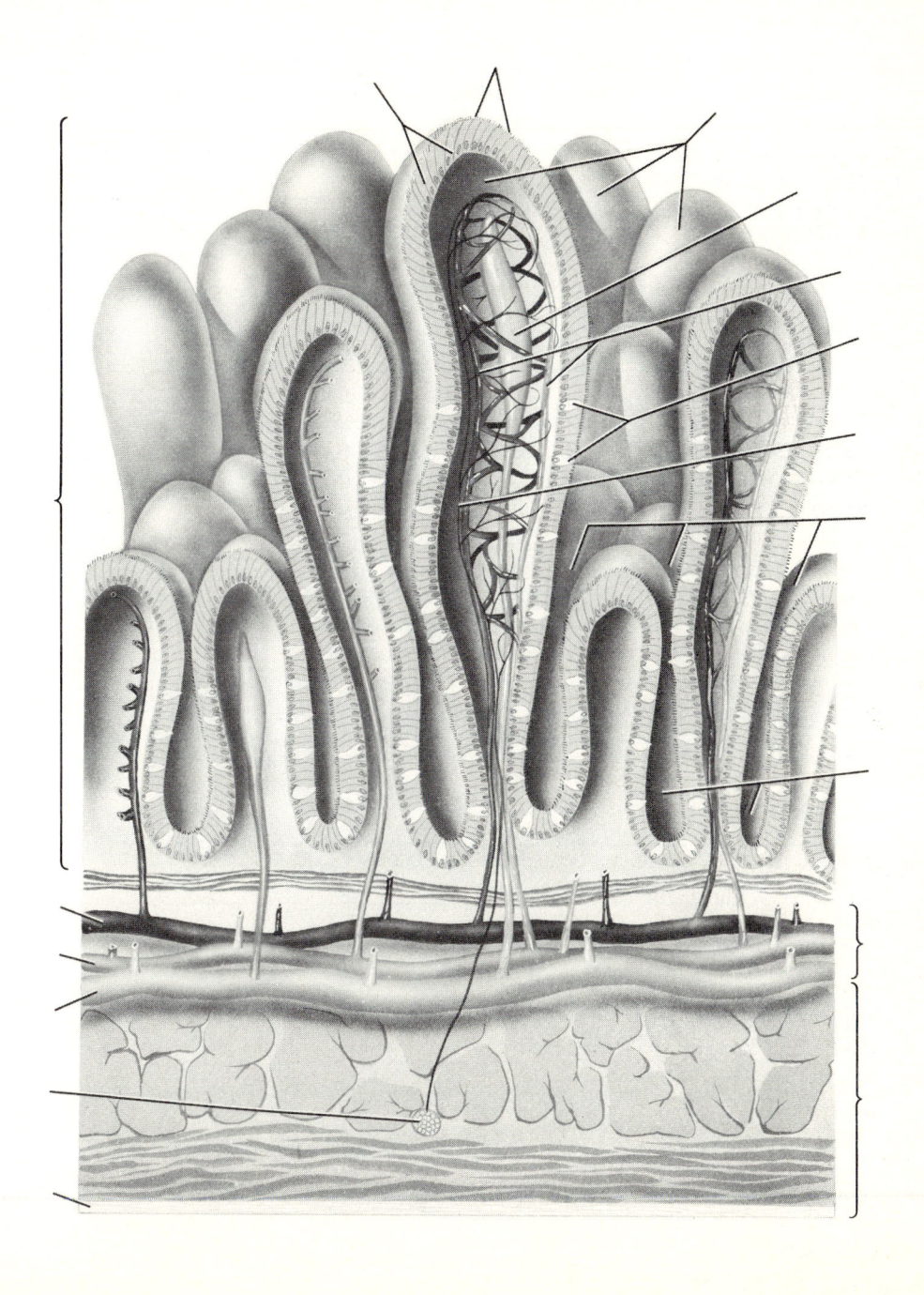

Absorption takes place through the villi. Glucose and amino acids are absorbed by active transport after combining with a carrier molecule. Fatty acids and monoglycerides are absorbed from micelles into the epithelial cells of the intestinal lining. There they are reassembled into triglycerides and packaged as chylomicrons.

(1) What is active transport? _____
(2) What is a chylomicaon? _____
(3) Where do chylomicrons go from the epithelial cells of the villi? _____

TEST YOURSELF

43. A chylomicron is   a. a protein encased fat globule   b. the form in which fat passes into the lacteal of the villus   c. a type of carrier molecule active in glucose absorption   d. a type of bile salt   e. answers (a) and (b) only are correct

> LO 27: Trace the chyme through the large intestine and out of the body as feces.

As the chyme moves slowly through the large intestine water and sodium is absorbed from it. The chyme is fashioned into feces which are expelled from the rectum via the anus.

(1) The solid portion of feces consists of _____

(2) What happens to the cellulose component of the chyme? _____

44. The solid portion of feces consists of   a. live and dead bacteria   b. cellulose   c. dead cells, salts, and bile pigments   d. answers (a),(b) and (c) are correct   e. answers (a) and (c) only are correct

> LO 28: Label the parts of the large intestine on a diagram.

The large intestine consists of the cecum and appendix, the ascending, transverse, descending and sigmoid colon, and the rectum which terminates in the anus.

(1) On a separate sheet of paper draw and label a diagram of the large intestine.
(2) What are haustra?_____
(3) What are taeniae?_____

TEST YOURSELF

45. Which part of the large intestine receives chyme directly from the ileum a. sigmoid colon b. cecum c. ascending colon d. descending colon e. transverse colon

> LO 29: Describe three functions of the large intestine, and give its common disorders.

The functions of the large intestine include absorption of sodium and water, incubation of bacteria, and formation and elimination of feces.

(1) Explain the mutualistic relationship we maintain with some of the bacteria that inhabit the large intestine_____

(2) Give the causes and physiologic basis for
   (a) diarrhea_____

   (b) constipation_____
(3) What is the suggested cause of cancer of the colon?
_____

TEST YOURSELF

46. Bacteria that inhabit the large intestine a. help break down cellulose b. produce certain vitamins c. may cause cancer of the colon d. answers (a), (b) and (c) are correct e. answers (a) and (b) only

ANSWERS TO TEST YOURSELF QUESTIONS

1.b	2.a	3.e	4.c	5.b	6.d	7.a	8.b	9.a
10.a	11.c	12.e	13.d	14.c	15.b	16.e	17.b
18.a	19.c	20.e	21.c	22.b	23.d	24.b	25.a
26.d	27.d	28.c	29.c	30.d	31.b	32.e	33.e
34.c	35.e	36.e	37.c	38.a	39.c	40.a	41.c
42.b	43.e	44.d	45.b	46.e

POST TEST

1. (LO 1) From the stomach chyme passes into the   a. gallbladder   b. esophagus   c. duodenum   d. jejunum   e. ileum

2. (LO 2) The process of expelling undigested food from the body, is   a. digestion   b. absorption   c. elimination   d. excretion   e. transport

3. (LO 3) The myenteric plexus is located   a. within the submucosa   b. between circular and longitudinal sublayers of the muscularis   c. lamina propria of the mucosa   d. outer connective tissue layer   e. none of these answers is correct

4. (LO 4) The membrane that suspends the stomach and duodenum from the liver is the   a. lesser omentum   b. greater omentum   c. parietal peritoneum   d. fatty apron   e. mesentery

5. (LO 5) Carbohydrate digestion begins in the   a. stomach   b. pancreas   c. mouth   d. salivary glands   e. duodenum

6. (LO 6) Fungiform papillae   a. are seen as red dots on the tip of the tongue   b. contain taste buds   c. are mushroom shaped   d. are threadlike   e. answers (a),(b) and (c) are correct

10. (LO 10) The bolus enters which portion of the stomach   a. corpus   b. fundus   c. antrum   d. pylorus   e. greater curvature

11. (LO 11) The chief cells of the gastric glands   a. are filled with granules   b. produce pepsinogen   c. produce intrinsic factor   d. produce hydrochloric acid   e. answers (a) and (b) only are correct

12. (LO 12) What happens to the food in the stomach   a. protein digestion begins   b. mechanical digestion takes place   c. lipids are emulsified   d. answers (a),(b) and (c) are correct   e. answers (a) and (b) only are correct

13. (LO 13) Peptic ulcers most often occur in the
    a. stomach   b. esophagus   c. duodenum   d. ileum
    e. jejunum

14. (LO 14) Which of the following occurs in the duodenum   a. receives bile from the liver   b. receives pancreatic juice from the pancreas   c. absorption of nutrients   d. enzyme release from brush border   e. all of the preceding answers are correct

15. (LO 15) The brush border of the small intestine is a consequence of the   a. plicae circulares   b. villi   c. taenia   d. microvilli   e. haustra

16. (LO 16) When acidic chyme comes into contact with the mucosa of the duodenum it stimulates release of   a. secretin   b. CCK   c. gastrin   d. pepsin   e. maltase

17. (LO 17) Trypsin acts to break down   a. cholesterol   b. maltose   c. dextrin   d. protein   e. triglycerides

18. (LO 18) Nutrients absorbed from the intestine are inventoried by   a. pancreas   b. liver   c. gallbladder   d. hepatic portal vein   e. villi

19. (LO 19) Hepatic sinusoids receive blood from branches of the   a. hepatic vein   b. hepatic arteries   c. canaliculi   d. central veins   e. answers (a) and (b) are correct

20. (LO 20) Bile salts are made from   a. billirubin   b. bile pigments   c. cholesterol   d. micelles   e. answers (a) and (b) are correct

21. (LO 21) The gallbladder contracts in response to   a. secretin   b. CCK   c. gastrin   d. acid chyme in the duodenum   e. cholesterol

22. (LO 22, 23, 24) Polypeptides are converted to dipeptides by   a. trypsin   b. lipase   c. amylase   d. maltase   e. peptidase

23. (LO 23) Lipid digestion takes place almost entirely within the   a. stomach   b. pancreas   c. duodenum   d. liver   e. gallbladder

24. (LO 25) Fat is absorbed mainly by the   a. lacteals of the villi   b. capillaries within the villi   c. tiny vein within each villus   d. tiny artery within each villus   e. sinusoids of the villi

25. (LO 26) Glucose and amino acids are absorbed mainly by   a. diffusion   b. conversion to chylomicrons   c. conversion to micelles   d. active transport   e. lacteals

26. (LO 27, 28) From the sigmoid colon feces enters the   a. appendix   b. ascending colon   c. descending colon   d. anus   e. rectum

27. (LO 29) Which of the following are functions of the large intestine   a. absorption of water   b. incubation of bacteria   c. formation of feces   d. answers (a), (b) and (c) are correct   e. answers (a) and (b) only are correct

ANSWERS TO POST TEST QUESTIONS

1.c	2.c	3.b	4.a	5.c	6.e	7.a	8.d	9.b
10.a	11.e	12.e	13.c	14.e	15.d	16.a	17.d
18.b	19.b	20.c	21.b	22.a	23.c	24.a	25.d
26.e	27.d

12 UTILIZATION OF NUTRIENTS
Metabolism

PRETEST

1. The building or synthetic aspect of metabolism is known as _____
2. _____ is an almost universal solvent and is used to transport materials in the body
3. _____ is a mineral required as a component of the thyroid gland
4. Vitamin _____ is a water soluble vitamin needed for collagen synthesis
5. The fat soluble vitamins are ___, ___, ___, and ___
6. Most carbohydrates are ingested as the polysaccharides: _____ and _____
7. In glycogenesis the liver packages _____ as _____ for storage
8. Glucose is utilized by the cells mainly as _____
9. The term _____ means to produce glucose from other types of molecules
10. After maximum amounts of glycogen have been stored excess glucose molecules are converted to _____ _____ and _____ and stored as _____
11. _____ is a molecule important in cellular energy storage
12. During glycolysis, glucose is converted to 2 _____ _____ molecules, 2 _____ and 4 _____
13. An oxygen debt may be incurred during _____ _____ when muscle cells respire _____
14. During the citric acid cycle, the fuel molecule is completely degraded to hydrogen and _____ _____

209

15. During aerobic respiration hydrogens are processed in the _____ _____ system
16. In aerobic respiration each glucose molecule may yield up to _____ ATPs
17. A fat that contains double bonds is a(n) _____ fat
18. _____ are used by cells as fuel and as components of cell membranes and steroid hormones
19. In general animal foods are rich in both _____ fat and _____
20. Low density lipoprotein contains mostly _____
21. Fatty acids are degraded in the liver cells by a process called _____ _____
22. Amino acids that must be included in the diet are known as _____ amino acids
23. Plant foods are often deficient in one or more _____ _____ acid
24. Excess amino acids are de-_____ and the amino groups are converted to _____ which is excreted by the _____
25. Kwashiorkor is caused by severe _____ _____
26. When the amount of nitrogen ingested and excreted in an adult are equal the person is said to be in _____ _____
27. A satiety center in the _____ acts to stop _____ behavior
28. The rate that energy is used by the body under resting conditions is the _____ _____
29. When energy input is greater than energy output a person will _____ _____
30. In order to lost weight calorie input must be _____ _____ energy output
31. When glycogen stores are depleted cells turn to _____ and _____ as sources of fuel

ANSWERS 1. anabolism (LO 1) 2. water (LO 2) 3. iodine (LO 3) 4. C (LO 4) 5. A,D,E,K (LO 5) 6. starch, cellulose (LO 6) 7. glucose, glycogen (LO 7) 8. fuel (LO 8) 9. gluconeogenesis (LO 9) 10. fatty acids, glycerol, fat (or triglycerides) (LO 10) 11. ATP (LO 11) 12. pyruvic acids, ATPs, hydrogens (LO 12) 13. strenuous exercise, anaerobically (LO 13) 14. carbon dioxide (LO 14) 15. electron transport (LO 15) 16. 38, (LO15)

16. 38, (LO 16)   17. unsaturated ,(LO 17)   18. fats (lipids) (LO 19)   19. saturated cholesterol (LO 18) 20. cholesterol (LO 20)   21.beta oxidation (LO 21) 22. essential (LO 22)   23. essential amino (LO 23) 24. aminated, urea, kidneys (LO 24)   25. protein defiency (LO 25)   26. nitrogen balance (LO 26)   27. hypothalamus, eating (LO 27)   28. basal metabolic (LO 28) 29. gain weight (LO 29)   30. less than (LO 30) 31. protein, fat (LO 31)

LO 1: Compare anabolism and catabolism

Metabolism may be divided into two broad categories: anabolism, or synthetic activities, and catabolism, or breaking down processes. Anabolism requires energy output; catabolism provides energy for physiologic processes.

1. Indicate whether each of the following activities is anabolic or catabolic:
    (a) synthesis of a protein from amino acids

    (b) manufacturing mitochondria_____
    (c) glycolysis_____
    (d) beta oxidation_____

TEST YOURSELF

Match:
1. anabolism___   a. conversion of glucose to pyruvic acid
2. catabolism___   b. cholesterol synthesis

LO 2: List and describe five ways in which the body utilizes water.

The body utilizes water as:
1._____
2._____
3._____
4._____
5._____

TEST YOURSELF

211

3. The body uses water a. to transport materials
   b. as a component of tissues and cells c. as a
   solvent d. to help maintain body temperature
   e. all of the preceding answers are correct

> LO 3: Identify minerals required by the body
> and give their functions in the body.

Minerals are inorganic nutrients ingested in the form of salts dissolved in food or water.

1. Give the function of each of the minerals listed below:
   (a) iron_____
   (b) iodine_____
   (c) calcium_____
   (d) phosphorus_____
   (e) sodium and chlorine_____

TEST YOURSELF
Match:

4. component of hemoglobin____   a. iodine
5. blood clotting; normal nerve   b. iron
   function____   c. fluorine
6. makes teeth resistant to decay____ d. calcium
                                      e. sodium

> LO 4: Give the actions and effect of defiency as well as major sources of the vitamins.

The vitamins are described in Table 12-2 of your textbook.

Give one important action and the effect of deficiency for the following vitamins.

	Action	Effect of deficiency
A		
D		
E		
K		

	Action	Effect of deficiency
Thiamin	_____	_____
Riboflavin	_____	_____
Niacin	_____	_____
Pantothenic acid	_____	_____
folic acid	_____	_____
B_{12}	_____	_____

TEST YOURSELF
Match:

7. constituent of Coenzyme A ___
8. essential for blood clotting ___
9. deficiency results in blindness ___
10. deficiency results in rickets ___

a. Vitamin D
b. Vitamin A
c. pantothenic acid
d. Vitamin K
e. niacin

> LO 5: Identify vitamins as water or fat-soluble and define hyper-vitaminosis.

(1) The water soluble vitamins are _____ and _____
    The fat soluble vitamins are ___, ___, ___, and ____
(2) Define hypervitaminosis _____
    _____

TEST YOURSELF

11. Large amounts of vitamin D   a. may cause symptoms of hypervitaminosis   b. are not harmful because vitamin D is water soluble   c. will simple result in excretion of the excess amount   d. will prevent rickets   e. answers (b), (c) and (d) are correct

> LO 6: List the principal types of carbohydrates ingested and their fate in the GI tract (that is what form are they in when absorbed.)

(1) Most carbohydrates are ingested as _____ and _____
(2) During digestion starch is hydrolyzed to _____
(3) Two important disaccharides in the diet are _____ and _____

213

(4) Most carbohydrates are absorbed in the form of
_____. Others are converted to
glucose in the _____

TEST YOURSELF

12. During digestion starch is hydrolyzed to   a. cellulose   b. sucrose   c. lactose   d. glycogen   e. glucose

> LO 7: Describe the role of the liver in regulating blood glucose level.

The liver removes excess glucose from the blood and converts it to glycogen for storage. As needed, glycogen is degraded and glucose returned to the blood. When glycogen stores are depleted liver cells convert amino acids and glycerol to glucose.

1. In what form does the liver store glucose? _____
2. Why is it necessary and important to regulate blood glucose level? _____
_____

3. What is the normal fasting level of glucose in the blood? _____

TEST YOURSELF

13. When excess glucose is present in the blood, the excess is   a. sent to the brain   b. stored as individual glucose molecules in the liver cells   c. converted to glycogen and stored in liver cells   d. converted to ATP which is stored in the liver cells   e. none of the preceding answers is correct

> LO 8: Describe the fate of glucose in the cells of the body.

Glucose is used as fuel in the cells. The energy released is used for doing physiologic work.

1. In the body cells glucose is utilized mainly as _____ in cellular respiration

2. In the presence of oxygen it is completely degraded to _____ and _____ _____ with the release of energy.
3. The energy is packaged in _____

TEST YOURSELF

14. In cellular respiration (when oxygen is present) glucose is completely converted to a. water, oxygen, and pyruvic acid b. ATP and hydrogen c. pyruvic acid and hydrogen d. glycogen, ATP, and carbon dioxide e. carbon dioxide, water and ATP

LO 9: Define the terms: Glycogenesis, glycogenolysis and gluconeogenesis.

1. Define:
   (a) glycogenesis _____
   (b) glycogenolysis _____
   (c) gluconeogenesis _____

TEST YOURSELF
Match:

15. convert glucose to glycogen
16. make glucose from other substances

a. glycogenolysis
b. glycogenesis
c. gluconeogenesis

LO 10: Tell what happens to excess glucose after the cells have stored maximum amounts of glycogen.

1. After maximum amounts of glycogen have been stored, excess glucose is converted to _____ _____ and _____
2. These compounds are then synthesized into _____ and sent to the _____ _____ for storage

TEST YOURSELF

17. When maximum amounts of glycogen have been stored, excess glucose is stored as a. protein b. amino acids c. single glucose molecules d. fat e. two of the preceding answers are correct

215

LO 11: *Describe ATP and give its role in cellular metabolism.*

ATP is composed of pentose, the base adenine, and three phosphate groups, two of which have high energy bonds. It serves as energy currency in cellular metabolism, being formed endergonically and breaking up exergonically in such a way that energy is transferred by phosphorylation and dephosphorylation.

1. Define:
   (a) endergonic_____
   (b) exergonic_____
2. Write an equation to show how ATP is produced from ADP:

3. Write an equation to show how ATP gives up its energy as it is converted back to ADP:

TEST YOURSELF

18. The formation of ATP from ADP   a. is an endergonic reaction, b. requires energy c. occurs when the cell needs energy for some cellular process d. answers (a),(b) and (c) are correct e. answers (a) and (b) only are correct

LO 12: *Summarize the reactions of glycolysis.*

The first several reactions of cellular respiration are referred to as glycolysis (or the Embden-Meyerhof pathway). Glycolysis might be summarized as follows:

$$\text{glucose} \longrightarrow 2 \text{ pyruvic acid} + 2 \text{ ATP} + 4 \text{ hydrogen}$$

1. Why must 2 ATPs be invested in the process of glycolysis?_____

2. Write a summary reaction for the production of PGAL from glucose:
   _____

3. Write a summary reaction for the formation of pyruvic acid from PGAL:
   _____

216

TEST YOURSELF

19. During glycolysis a.2 ATPs must be invested in the reactions taking place b. there is a net gain of 2 ATPs c. 2 PGALs are formed from one glucose d. 4 hydrogens are pulled off of one glucose molecule e. all of the preceding answers are correct

> LO 13: Discuss oxygen debt in relation to anaerobic respiration.

When sufficient oxygen is not available hydrogen is accepted by pyruvic acid which thereby becomes lactic acid. Extra oxygen is then needed to dispose of the lactic acid.

(1) What is an oxygen debt?_____
(2) What causes muscle fatigue?_____
(3) How does lactic acid help stimulate the body to obtain the needed oxygen to pay back the debt?
_____

TEST YOURSELF

20. When one engages in strenuous exercise a. an oxygen debt is incurred b. large amounts of pyruvic acid build up in the muscles c. pyruvic acid acidifies the blood so that respiration increases d. answers (a),(b) and (c) are correct e. none of the preceding answers is correct

> LO 14: Summarize the citric acid cycle in whatever degree of detail specified by your instructor.

In the citric acid (Krebs) cycle the fuel molecule is completely consumed. Its breakdown products are hydrogen and carbon dioxide. The hydrogen enters the electron transport system and eventually combines with oxygen to form water.

(1) Before pyruvic acid actually enters the citric acid cycle it is degraded to a 2-carbon compound with the release of_____

(2) The 2-carbon compound which we shall call acetyl combines with a special enzyme_____ to form_____
(3) Coenzyme A is synthesized from_____
(4) The acetyl compound combines with a 4-carbon compound to form a 6-carbon compound,_____
(5) How many turns of the citric acid cycle would be required to process one glucose molecule?_____ Why?_____

TEST YOURSELF
Match:

21. A 6-carbon compound___
22. synthesized from pantothenic acid
23. A 2-carbon compound___

a. pyruvic acid
b. citric acid
c. Coenzyme A
d. oxaloacetic acid
e. acetyl compound

LO 15: Describe the electron transport system and relate it to aerobic cellular respiration.

The electron transport system consists of the hydrogen acceptors NAD and FAD, plus a series of mitochondrial cytochromes. These compounds by a process of electron transfer extract energy in stepwise fashion from hydrogen when oxygen is available.

(1) Where do the reactions of the electron transport system take place?_____
(2) What is the final hydrogen (and electron) acceptor in the electron transport system?_____
(3) Why does the body require oxygen? Be specific.
(4) What happens to the electron system when oxygen is not available?_____
(5) How does cyanide poison the body?_____

TEST YOURSELF
24. For every two hydrogen atoms that pass through the electron transport system how many ATPs are produced:
   a. 1   b. 2   c. 3   d. 4   e. 6

218

LO 16: Compare aerobic and anaerobic respiration with respect to ATP production, hydrogen acceptor, and end products.

Fill in the chart:

	Anaerobic	Aerobic
Total ATP profit	_____	_____
End products	_____	_____
Final H Acceptor	_____	_____

TEST YOURSELF

Match:
25. Total ATP profits in aerobic respiration____
26. End products in aerobic respiration____

a. 2 ATP
b. 38 ATP
c. lactic acid, water
d. carbon dioxide, water
e. pyruvic acid, oxygen

LO 17: Distinguish between saturated and unsaturated fats and among monoglycerides, diglycerides, and triglycerides.

Saturated fats are fully loaded with hydrogens while unsaturated fats contain one (monounsaturated) or more (polyunsaturated) fatty acids to which additional hydrogens could be added. In monoglycerides, one fatty acid is attached to a glycerol molecule, in diglycerides, two fatty acids are attached to a glycerol molecule, and in triglycerides, three fatty acids are attached to a glycerol molecule.

(1) What type of bond distinguishes an unsaturated fatty acid?____
(2) Could there by a fat with 4 fatty acids attached to a glycerol molecule?_____Why, or why not?
_____

TEST YOURSELF

27. Which answer best describes a fat that has 2 fatty acids attached to a glycerol molecule, and in which one of those fatty acids has double bonds: a. diglyceride b. triglyceride c. polyunsaturated diglyceride d. monounsaturated triglyceride e. monounsaturated diglyceride

> LO 18: *Identify foods that are rich in saturated fats and cholesterol and those rich in polyunsaturated fats, and summarize the hypothesis that dietary intake of these contributes to development of atherosclerosis.*

(1) _____ fats are rich in saturated fat and cholesterol. _____ oils are rich in _____ and have no _____
(2) Ingestion of polyunsaturated fats appears to decrease blood cholesterol level, thus giving some protection against _____
(3) Ingestion of large amounts of cholesterol and saturated fats appears to increase the chances of developing _____

TEST YOURSELF

28. Some investigators believe that you can help protect yourself from developing atherosclerosis by
a. substituting plant oils for animal fat
b. substituting saturated for unsaturated fat
c. reducing intake of unsaturated fat   d. substituting cholesterol for saturated fat   e. substituting saturated fat for cholesterol

> LO 19: *Trace the fate of lipids from the time of absorption to their deposit and storage in adipose tissue, and tell how they are used in the body.*

Chylomicrons are absorbed into the lymphatic system and transported to the blood. From the blood chylomicrons are absorbed into fat cells.

List four functions of fat in the body:
1._____
2._____
3._____
4._____

TEST YOURSELF

29. Which of the following is NOT a function of lipid in the body a. used as fuel b. needed for oxidation by brain cells c. synthesis of cell membranes d. synthesis of bile salts e. synthesis of steroid hormones

> LO 20: Give the function of lipoprotein in transporting lipids, and describe the mobilization of fat from fat cells.

As needed fat is mobilized from fat cells. Triglycerides are hydrolyzed within the fat cells and free fatty acids are sent into the blood. They are transported in combination with plasma albumin., a protein in the blood.

(1) What is a lipoprotein?_____
(2) Is a chylomicron a lipoprotein?_____
(3) Why is fat especially important as fuel between meals?_____

TEST YOURSELF

30. Fat is normally found in the blood in a. combination with plasma albumin b. chylomicrons c. free floating fat cells d. answers (a) and (b) only are correct e. answers (a),(b) and (c) are correct

> LO 21: Summarize beta oxidation of fatty acids and discuss its significance.

Before fatty acids can be utilized as fuel they are degraded by beta oxidation to molecules of acetyl Coenzyme A.

(1) Beta oxidation takes place primarily in the_____ cells
(2) Why are some molecules of acetyl Coenzyme A converted to ketone bodies?_____
(3) When completely degraded a 6-carbon fatty acid generates_____ATPs

TEST YOURSELF

11. During beta oxidation fats are converted to molecules of a. acetyl coenzyme A   b. triglycaride   c. Pyruvic acid   d. ATP   e. acetone

> LO 22: Tell why essential amino acids must be included in the diet, and identify foods as sources of high or low quality protein.

Essential amino acids must be included in the diet because the body is not able to synthesize them from other nutrients.

(1) What is the fate of essential amino acids in the body?_____
(2) In general animal foods are sources of high quality protein because their proteins contain an appropriate distribution of_____

(3) Plant protein often lacks an appropriate balance of_____amino acids. Legumes are good sources of plant protein; cereal grains are not as good.

TEST YOURSELF
Match

32. corn_____   a. high quality protein
33. milk_____   b. low quality protein

> LO 23: Give three reasons why it is more difficult to obtain adequate amino acid amounts in a vegetarian diet, and tell how a nutritious vegetarian diet can be planned.

A nutritious vegetarian diet can be planned by using a variety of plant foods that complement each other with regard to amino acid content. Use of dairy products with plant foods greatly enhances essential amino acid intake.

(1) What animal food is often combined with each of the following plant foods to enhance its nutritional value:
  (a) macaroni and _____
  (b) cereal and _____
  (c) spaghetti and _____
(2) Give 3 reasons why it is more difficult to obtain adequate amino acid amounts from a vegetarian diet:
  1. _____
  2. _____
  3. _____

TEST YOURSELF

34. Plant foods a. contain protein that is encased within indigestible cellulose plant walls b. contain a lower percentage of protein than animal foods c. are often deficient in one or more essential amino acids d. answers (a),(b) and (c) are correct e. answers (a) and (c) only are correct

> LO 24: Give the fate and actions of amino acids and proteins in the body, and describe the fate of excess amino acids.

Amino acids are used to make proteins which are essential as structural proteins and as enzymes, as well as for other specific purposes such as manufacture of hemoglobin in red blood cells.

(1) Name one use of a structural protein in the body _____
(2) Name an enzyme _____
(3) Excess amino acids are _____ and the remaining keto acid is converted to _____ for use as fuel or to lipid for storage as _____

TEST YOURSELF

35. Excess amino acids are   a. stored in the liver
    b. converted to protein and stored in liver and
    muscle cells   c. immediately converted to ATP
    d. deaminated, then converted to carbohydrate or
    lipid   e. excreted via the kidneys

> LO 25: List effects of protein deficiency
> and describe the disease KWASHIORKOR.

Protein deficiency results in physical and mental retardation, lowered resistance to disease, and increased risk of death from ordinarily mild diseases. Extreme protein deficiency results in kwashiorkor.

(1) Can an unborn child be affected if its mother suffers from protein deficiency?_____How?

(2) Explain how children often develop kwashiorkor

(3) What are some of the symptoms of this disease?

TEST YOURSELF

35. Which of the following might result from severe
    protein deficiency?   a. death from measles
    b. smaller, less developed brain   c. stunted
    physical growth   d. answers (a),(b) and (c) are
    correct   e. answers (b) and (c) only are correct

> LO 26: Tell what is meant by nitrogen balance.

When the amount of nitrogen ingested daily by an adult equals the amount excreted the body is in nitrogen balance.

(1) The amount of nitrogen in the urine reflects the amount of breakdown of _____
(2) How is the situation different in a growing child?

TEST YOURSELF

37. In a growing child intake of protein nitrogen generally   a. equals nitrogen excretion   b. exceeds nitrogen excretion   c. is less than nitrogen excretion   d. just meets growth needs and no nitrogen is excreted   e. none of the preceding answers is correct

> LO 27: Describe mechanisms that regulate food intake.

Food intake is regulated by a feeding center and by a satiety center in the hypothalamus. These in turn appear to be regulated by the concentration of glucose and amino acids in the blood.

(1) When the feeding center is damaged in an experimental animal what happens? _____

(2) What happens when the satiety center is destroyed in an experimental animal? _____

TEST YOURSELF

38. When the level of glucose in the blood increases, the   a. feeding center is stimulated   b. satiety center is inhibited   c. satiety center is stimulated   d. eating behavior increases   e. two of the preceding answers are correct

> LO 28: Give the meaning of basal metabolic rate and describe how it is measured. Compare with total metabolic rate.

Basal metabolic rate is the amount of energy utilized by the body under resting conditions. It is measured by determining the amount of oxygen consumed during a given period. Total metabolic rate is the sum of BMR and the energy needed for carrying on all of one's daily activities.

(1) What is meant by basal conditions? _____

(2) A person with a BMR that is 30% above normal would have a BMR of _____
(3) A patient with a BMR that is 10% below normal would have a BMR of _____

TEST YOURSELF

39. Which of the following would probably have the highest total metabolic rate   a. a college professor   b. a secretary   c. construction worker   d. student   e. executive

> LO 29: Give the basic energy equation for maintaining body weight, and tell what happens if it is altered in either direction.

Body weight remains constant when energy (calorie) input = energy output.

(1) What happens when: calorie input $<$ energy output
_____

(2) What happens when: calorie input $>$ energy output
_____

TEST YOURSELF
Match:

40. gain weight when _____   a. E input = E output
41. maintain weight when _____   b. E input $<$ E output
                                   c. E input $>$ C output

> LO 30: Give the effects of obesity on health, its causes, and its cure based upon energy considerations. Describe the most effective reducing diet.

Obesity stresses the heart and increases the risk of dying from heart disease, cerebral hemorrhage, and diabetes. The basic cause of obesity is eating too much in relation to energy expended in daily activities.

(1) Why is it a disservice to overfeed a baby?

(2) Describe the most effective reducing diet

TEST YOURSELF

42. The best reducing diet is one that is   a. high in protein with almost no carbohydriates   b. high in fat with very little protein or carboyhydrate   c. high carbohydrate   d. water supplemented with vitamin pills   e. a normal proportion of fats, carbohydrates, and proteins

> LO 31: Summarize the metabolic changes that take place during starvation and describe marasmus.

During starvation the cells turn to protein and fat as sources of fuel. After several days of fasting brain cells activate enzymes needed to utilize ketone bodies for fuel.

(1) When the body depends upon fat for fuel absnormal amounts of ketone bodies gbuild up and may cause

(2) Describe marasmus

TEST YOURSELF

43. During starvation   a. the brain begins to utilize ketone bodies as fuel   b. muscle cells depend mainly on glycogen as fuel   c. other cells deplete their own structural protein until they have consumed themselves   d. answers (a),(b) and (c) are correct   e. answers (b) and (c) only are correct

ANSWERS TO TEST YOURSELF QUESTIONS

1.b	2.a	3.e	4.b	5.d	6.c	7.c	8.d	9.b
10.a	11.a	12.e	13.c	14.e	15.b	16.c	17.d
18.e	19.e	20.a	21.b	22.c	23.e	24.c	25.b
26.d	27.e	28.a	29.b	30.d	31.a	32.b	33.a
34.d	35.d	36.d	37.b	38.c	39.c	40.c	41.a
42.e	43.a

POST TEST

1. (LO 1) Protein synthesis is an example of a. an anabolic process b. a catabolic process c. a mechanism used during periods of starvation d. a mechanism for triggering the hunger center e. two of the precedint are correct

2. (LO 2) About 60 percent of the body by weight is a. glucose b. amino acids c. water d. glycogen e. mineral

3. (LO 3,4) The principal positive ion (cation) in interstitial fluid is a. chlorine b. iron c. thiamin d. sodium e. riboflavin

4. (LO 3,4) Niacin is used to make a. bone b. thyroid hormone c. hemoglobin d. Coenzyme A e. NAD

5. (LO 5) Which of the following vitamins is/are fat-soluble a. thiamin b. vitamin A c. vitamin C d. riboflavin e. two of the preceding answers are correct

6. (LO 6) Most carbohydrates are absorbed in the form of a. glycogen b. glucose b. maltose d. dextrin e. amino acids

7. (LO 7,9) The process of converting glycogen to glucose is a. gluconeogenesis b. glycogenesis c. glycogenolysis d. beta oxidation e. electron transport

8. (LO 8) Cells use glucose primarily as a. fuel b. building blocks for making cell membranes c. constituents of proteins d. a storage form of energy e. components of enzymes

9. (LO 10) Excessive amounts of carbohydrate in the diet end up as a. enzymes b. fat c. amino acids d. essential amino acids e. stored maltose

10. (LO 11,12) During glycolysis energy from the fuel molecule is stored in a. ATP b. coenzyme A c. triglyceride d. glycogen e. fatty acids

11. (LO 12,13) When sufficient oxygen is not available hydrogen is accepted by a. pyruvic acid b. coenzyme A c. PGAL . ethyl alcohol e. citric acid

12. (LO 14) How many turns of the citric acid cycle are/is required to process one glucose molecule a.1 b.2 c.3 d.4 e.6

13. (LO 15) Which of the following is/are electron acceptors in the electron transport system? a. oxygen b. NAD c. FAD d. cytochromes e. all of the preceding answers are correct

14. (LO 16) The final hydrogen acceptor in aerobic respiration is a. pyruvic acid b. lactic acid c. NAD d. cytochrome B e. molecular oxygen

15. (LO 17,18) Which of the following would be rich in cholesterol and saturated fat a. beef b. corn c. soybeans d. wheat e. answers (b),(c) and (d) are correct

16. (LO 19,20) Absorbed fat is transported to the fat depots a. as ketone bodies b. in combination with serum albumin c. in chylomicrons d. as lipoglycogen e. mainly as free triglycerides

17. (LO 21) The process by which fatty acids are degraded to acetyl Coenzyme A is called a. glycolysis b. citric acid cycle c. electron transport d. beta oxidation e. ketone body formation

18. (LO 22) Which of the following contain high quality protein? a. milk b. corn c. eggs d. peanuts e. answers (a) and (c) are correct

19. (LO 23) A nutritious vegetarian diet would include a. mainly cereal grains b. a variety of plant foods eaten at the same meal and supplemented with dairy products or fish c. any one of the legumes d. corn or rice as the staple food e. answers (a) and (d) are correct

20. (LO 24) Amino acids are used a. to make enzymes b. to make protein such as hemoglobin c. to make structural proteins d. as fuel e. all of the preceding answers are correct

21. (LO 25,26) When the body is in negative nitrogen balance a. the body excretes less nitrogen than it takes in b. very little body protein must be degraded c. needed proteins cannot be synthesized d. excess amino acids are stored in the liver cells e. answers (a) and (b) are correct

22. (LO 27) Food intake is thought to be regulated by a. centers in the spinal cord b. level of glucose in the blood c. level of hemoglobin in the blood d. needs of the fat cells e. answers (a) and (b) are correct

23. (LO 28) The energy you comsume to support your metabolism and your daily activities is your a. total metabolic rate b. BMR c. satiety level d. nitrogen balance e. energy balance

24. (LO 29,30) When a person is losing weight a. his energy input is less than his energy output b. a portion of his energy needs is being met by mobilizing fat c. energy input is equal to energy output d. essential amino acids are stored more rapidly than usual e. answers (a) and (b) only are correct

25 (LO 31) John's brain cells are utilizing ketone bodies for fuel. John a. has just eaten a big meal b. has rickets c. must have kwashiorkor d. is suffering from starvation e. is probably between meals

ANSWERS TO POST TEST QUESTIONS

1.a   2.c   3.d   4.e   5.b   6.b   7.c   8.a   9.b
10.a 11.a 12.b 13.e 14.e 15.a 16.c 17.d
18.e 19.b 20.e 21.c 22.b 23.a 24.e 25.d

13 THE CARDIOVASCULAR SYSTEM

PRETEST

1. The cardiovascular system transports _nutrients_, _____, and _____ to the cells of the body
2. The fluid portion of the blood is called _____; the cellular elements of the blood are the _____, _____, and _____
3. Red blood cells develop in the _____
4. Red blood cell production is regulated by a hormone called _____
5. A deficiency of hemoglobin is termed _____
6. Macrophages develop from blood cells called _____
7. In acute _____ the majority of white cells in the blood are immature forms
8. Megakaryocytes give rise to _____
9. During clotting thrombin catalyzes the conversion of _____ to _____
10. Very elastic walls are characteristic of what type of blood vessel? _____
11. From the right ventricle blood passes into the _____ arteries
12. Three large arteries that branch from the aortic arch are the _____, _____ and _____
13. The vertebral arteries bring blood to the _____
14. The mitral valve is located between the _____ and _____
15. The AV bundle consists of specialized muscle fibers called _____ fibers
16. Cardiac muscle cells are separated at their ends by dense bands called _____

17. The period of cardiac contraction is called _____
18. The P wave of the ECG represents the spread of an impulse over the _____
19. When signalled by its _____ nerves the heart slows
20. _____ Law of the heart tells us that the heart pumps all of the blood delivered to it within physiologic limits
21. As the radius of a blood vessel decreases resistance to blood flow _____
    (increases, decreases)
22. Arterial baroreceptors are sensitive to changes in blood _____
23. The alternate expansion and recoil of an artery is the _____
24. In _____ plaques develop within the inner layer of the arterial walls
25. The condition in which increased demand upon the heart results in chest pain is known as _____

ANSWERS: 1. nutrients, oxygen, hormones (LO 1)
2. plasma; red blood cells, white blood cells, platelets (LO 2) 3. bone marrow (LO 3) 4. erythropoetin (LO 4) 5. anemia (LO 5) 6. monocytes (LO 6) 7. leukemia (LO 7) 8. platelets (LO 8) 9. fibrinogen to fibrin (LO 9) 10. arteries (LO 10) 11. pulmonary (LO 11) 12. brachiocephalic, left common carotid, left subclavian (LO 12) 13. brain (LO 13) 14. left atrium, left ventricle (LO 14) 15. Purkinje (LO 15) 16. intercalated discs (LO 16) 17. systole (LO 17) 18. atria (LO 18) 19. parasympathetic (LO 19) 20. Starling's (LO 20) 21. increases (LO 21) 22. pressure (LO 22)
23. pulse (LO 23) 24. atherosclerosis (LO 24)
25. angina (LO 25)

> LO 1: List five functions of the cardiovascular system.

The cardiovascular system transports:
1. _____
2. _____
3. _____
4. _____
Other functions:
5. _____

TEST YOURSELF

1. Which of the following is NOT a function of the cardiovascular system: a. transports hormones   b. helps regulate body temperature   c. produces oxygen   d. transports nutrients   e. defends body against disease causing organisms?

> *LO 2: List the principal components of blood and briefly describe each.*

Blood consists of a fluid called plasma in which red blood cells, white blood cells, and platelets are suspended.

(1) Write a few key words of description to help you remember the components of blood:
   (a) plasma_____
   (b) red blood cells_____
   (c) white blood cells_____
   (d) platelets_____
(2) List the principal components of plasma:_____
_____

TEST YOURSELF
Match:

2. fluid portion of blood___   a. red blood cells
3. percent known as the hema-   b. white blood cells
   tocrit___   c. platelets
4. important in clotting___   d. plasma

> *LO 3: Trace a red blood cell through its life cycle from stem cell to maturity and describe the structure and function of a mature red blood cell.*

Stem cells in the bone marrow multiply to give rise to immature red blood cells. These synthesize hemoglobin, expel their nuclei, and enter the circulation. The mature red blood cell is a flexible biconcave disc lacking a nucleus and other cellular organelles.

(1) What is the function of the mature red blood cell?
_____

(2) Briefly describe:
   (a) stem cell _____
   (b) pronormoblast _____
   (c) reticulocyte _____
   (d) hemoglobin _____
   (e) ferritin _____

TEST YOURSELF

Place the stages of red blood cells into proper sequence
5. _____   a. reticulocyte
6. _____   b. pronormoblast
7. _____   c. mature erythrocyte
8. _____   d. uncommitted stem cell

> LO 4: Describe the role of erythropoietin in regulation of red blood cell production.

Hypoxia is thought to trigger release of REF and thus production of erythropoietin which stimulates the stem cells to step up red blood cell production.

(1) What is the source of REF? _____
(2) Where is erythropoietin formed? _____
    From what? _____

TEST YOURSELF

9. Erythropoietin acts upon a. REF b. kidney cells c. a globulin in the blood d. liver cells e. stem cells in the bone marrow

> LO 5: Give three general causes of anemia and describe specific types of anemia discussed (e.g. sickle cell anemia and pernicious anemia).

Anemia may be caused by (1) blood loss, (2) decreased production of red cells, or (3) increased rate of red cell destruction.

(1) Define anemia _____
_____

(2) Identify the cause of each of the following types
of anemia. (Use the categories listed on the
preceding page.)
(a) Aplastic anemia_____
(b) Iron deficiency_____
(c) Sickle cell_____
(d) Hemolytic anemia_____
(e) Pernicious anemia_____

TEST YOURSELF
Match:

10. Hemolytic anemia___   a. iron deficiency
11. Aplastic anemia___   b. sickle cell is an exam-
12. Pale (hypochromic), micro-   ple
    cytic red cells are   c. may be caused by
    characteristic___   excessive exposure to
                                    radiation

LO 6: Describe (and identify in diagrams or photomicrographs) five types of white blood cells, and give the functions of each type.

Three types of <u>granular</u> leukocytes are neutrophils, eosinophils, and basophils. Lymphocytes and monocytes are <u>agranular</u> leukocytes.

Complete the table:

	Description	Draw one	Function
Neutrophil			
Eosinophil			
Basophil			
Lymphocyte			
Monocyte			

TEST YOURSELF
Match:

13. become mast cells___
14. become macrophages___
15. granules stain bright orange___

a. basophils
b. lymphocytes
c. monocytes
d. neutrophils
e. eosinophils

> LO 7: Describe leukemia, telling what is known about its cause and describing its pathologic changes.

Leukemia is a form of cancer in which stem cells in the bone marrow produce excessive numbers of abnormal white blood cells.

(1) What factors are thought to "cause" leukemia?
_____

(2) Why would a victim of leukemia tend to be anemic?
_____

(3) Why do many victims of leukemia die of pneumonia or other infections?_____
_____

TEST YOURSELF

16. Which of the following would be characteristic of a leukemia victim   a. leukocytosis   b. anemia   c. low resistance to infection   d. answers (a),(b) and (c) are correct   e. answers (b) and (c) only are correct

> LO 8: Give the origin and describe the structure of a platelet.

Platelets are bits of cytoplasm pinched off from large megakaryocytes in the bone marrow.

(1) What are megakaryocytes?_____
(2) What are thrombocytes?_____
(3) Do platelets have nuclei?_____ Explain,
_____

TEST YOURSELF

17. Platelets a. are also called thrombocytes b. are more numerous than red blood cells c. are produced in the liver d. answers (a),(b) and (c) are correct e. answers (a) and (b) only are correct

> LO 9: Describe the formation of a platelet plug and summarize <u>the chemical</u> events of of blood clotting.

When a blood vessel is cut platelets gather to form a temporary clot. At the same time a permanent clot begins to form via a series of chemical reactions that begins when the damaged vessel activates the first clotting factor.

(1) At the end of the sequence of clotting reactions, activated factor X in the presence of calcium and phospholipids catalyzes the conversion of_____ to_____. Then thrombin catalyzes the conversion of_____ to_____ which forms the long threads of the clot

(2) Write the two final reactions just summarized:

(3) In the intrinsic pathway the series of clotting reactions are triggered when inactive factor_____ (also called_____ _____) is activated by the damaged_____, probably by exposure of the blood to_____

(4) In the extrinsic pathway the damaged tissue releases factor III, also called_____

(5) Why is vitamin K needed for normal blood clotting?
_____

TEST YOURSELF

Match:
18. converts prothrombin to   a. factor XII
    thrombin___   b. thromboplastin
19. converts fibrinogen to   c. thrombin
    fibrin___   d. factor X
20. triggers sequence of reactions   e. fibrin
    in intrinsic pathway___

> *LO 10: Compare the structures and functions of the different types of blood vessels including arteries, arterioles, capillaries, sinusoids, venules, and veins.*

A blood vessel wall consists of (1) <u>tunica intima</u>, the endothelial lining, (2) <u>tunica media</u> consisting of smooth muscle cells that control the diameter of the vessel and in large arteries of elastic connective tissue, and (3) <u>tunica adventitia</u>, fibrous connective tissue that contains nerves and blood vessels that supply the wall.

(1) Draw a diagram showing how the vessels listed in LO 10 would be arranged in a circulatory pathway

(2) Briefly describe and give the function of each type of vessel:

	Description	Function
Artery		
Arteriole		
Capillary		
Sinusoid		
Venule		
Vein		

<u>TEST YOURSELF</u>

Match:
21. Exchange of materials between blood and body tissues ___
22. have valves ___
23. very important in regulating blood supply to a particular tissue ___

a. capillaries
b. arteries
c. arterioles
d. veins
e. venules

> LO 11: Describe the general pattern of blood flow through the pulmonary and systemic circulation by tracing a drop of blood through each major vessel and heart chamber.

(1) A drop of blood entering the right atrium would flow in sequence through the following vessels:

right atrium ⟶ right _____ ⟶ pulmonary _____ ⟶ pulmonary arterioles ⟶ p.capillaries ⟶ pulmonary venules ⟶ left atrium ⟶ _____

(2) From the left ventricle blood would pass into the systemic circulation via the largest artery in the body, the _____. From the systemic circulation blood returns to the right atrium via the _____ or _____.

(3) Label the structures in the diagram:

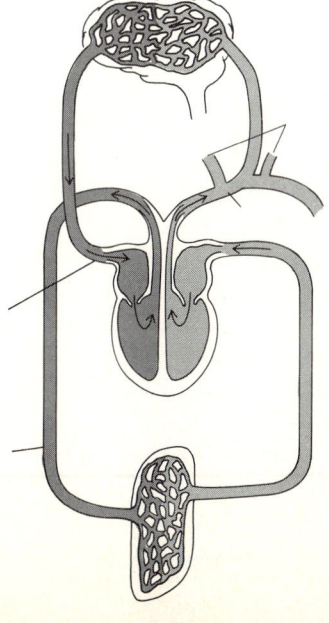

TEST YOURSELF

Indicate the vessel or chamber through which blood would pass through NEXT in sequence after flowing through:

24. left atrium___   a. aorta
25. inferior vena cava___   b. pulmonary vein
26. right ventricle___   c. pulmonary artery
                            d. left ventricle
                            e. right atrium

> LO 12: Identify the main portions of the aorta and list its principal branches.

The ascending aorta gives rise to the coronary arteries. The aortic arch gives rise to the brachiocephalic, left common carotid, and left subclavian. The thoracic portion of the descending aorta gives rise to the intercostal arteries, subcostal arteries, bronchials, and esophageals. The abdominal aorta gives rise to the celiac, superior mesenteric, suprarenals, renals, ovarian or testicular, inferior mesenteric, and common iliac arteries as well as to parietal branches including inferior phrenic, lumbar, and middle sacral arteries.

(1) Label the main portion of the aorta and label the principal branches shown on the diagram; located on following page.

(2) On a separate sheet of paper, draw a diagram of the aorta and label each of its divisions. Show the origin of the coronary arteries, brachiocephalic, left common carotid, left subclavian, celiac, renal, and common iliac arteries.

TEST YOURSELF

Which part of the aorta does each of the following arteries originate from?
27. left common carotid___   a. Ascending aorta
28. coronary arteries___   b. Aortic arch
29. inferior mesenteric___   c. Thoracic aorta
30. celiac___   d. Abdominal aorta
31. intercostal arteries___

LO 13: Trace the blood through the principal vessels of the coronary circulation: the cerebral circulation: the hepatic portal system. Identify the principal veins that deliver blood to the inferior and superior vena cava.

(1) Coronary circulation: The left coronary artery passes under the left atrium and divides into the anterior descending branch which supplies the walls of both _____, and the _____ branch which serves the walls of the left ventricle and left atrium. The right coronary artery divides into the posterior descending branch which supplies both ventricles and the _____ branch which supplies right atrium and ventricle. Capillaries empty into coronary veins which join to form the coronary sinus. Blood from the coronary sinus is returned to the ____ _____ of the heart.

(2) Cerebral circulation: Subclavian arteries ⟶ vertebral arteries ⟶ basilar a ⟶ rt. and left posterior cerebral a.

Internal carotid a. ⟶ anterior cerebral a. and middle cerebral a.

The anterior and middle and posterior cerebral arteries are joined by small communicating arteries. The arterial anastomosis at the base of the brain is the _____ ___ _____ .

(3) Hepatic portal system: Intestinal capillaries ⟶ superior mesenteric vein ⟶ hepatic portal v. ⟶ hepatic sinusoids ⟶ hepatic veins ⟶ inferior vena cava. Hepatic sinusoids also receive blood from the _____ _____

(4) Label the diagram illustrating the circulation of blood through the hepatic portal system.

(See Figure on following page)

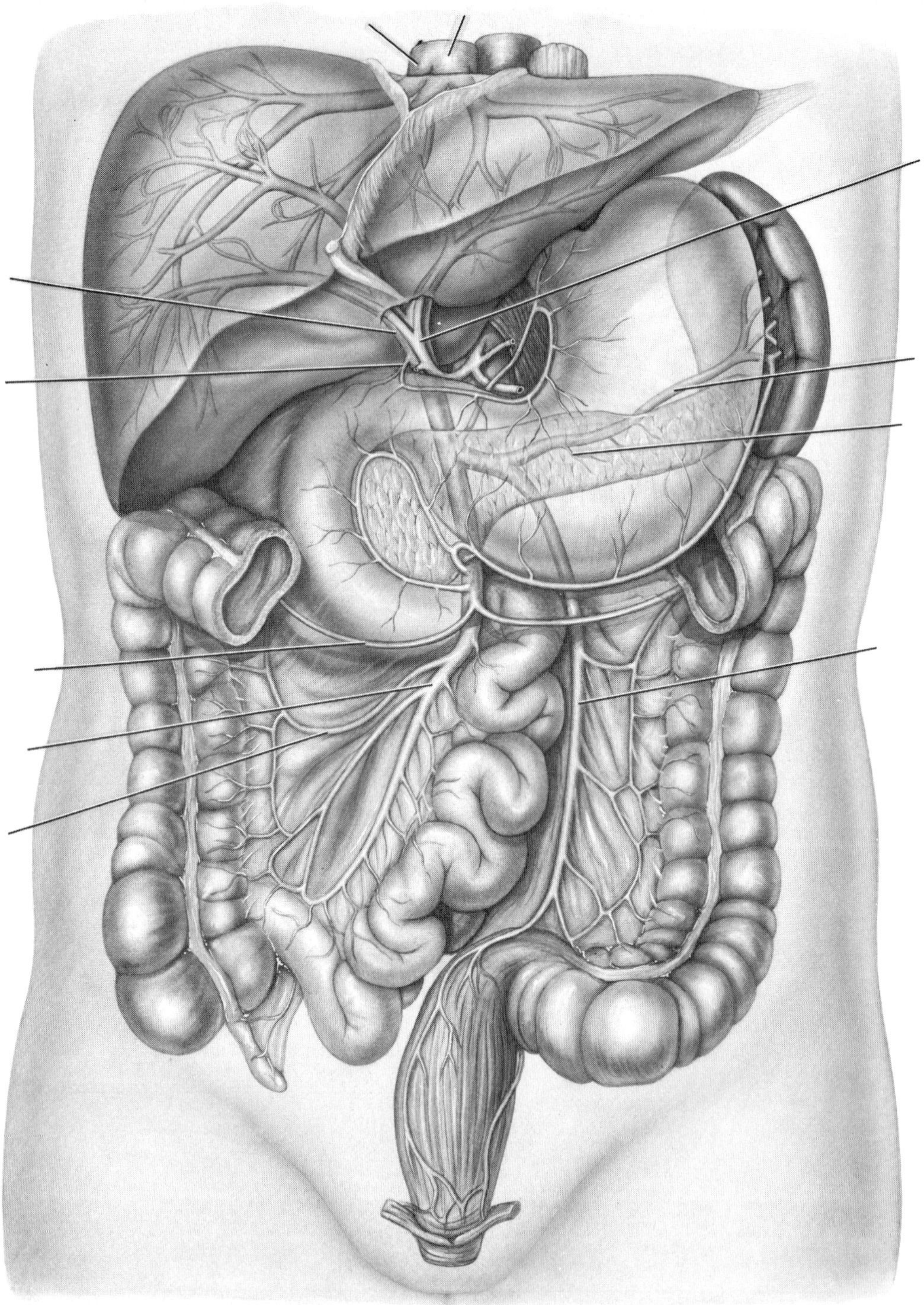

(5) Label the diagram illustrating the principal veins that empty into the superior and interior vena cavae.

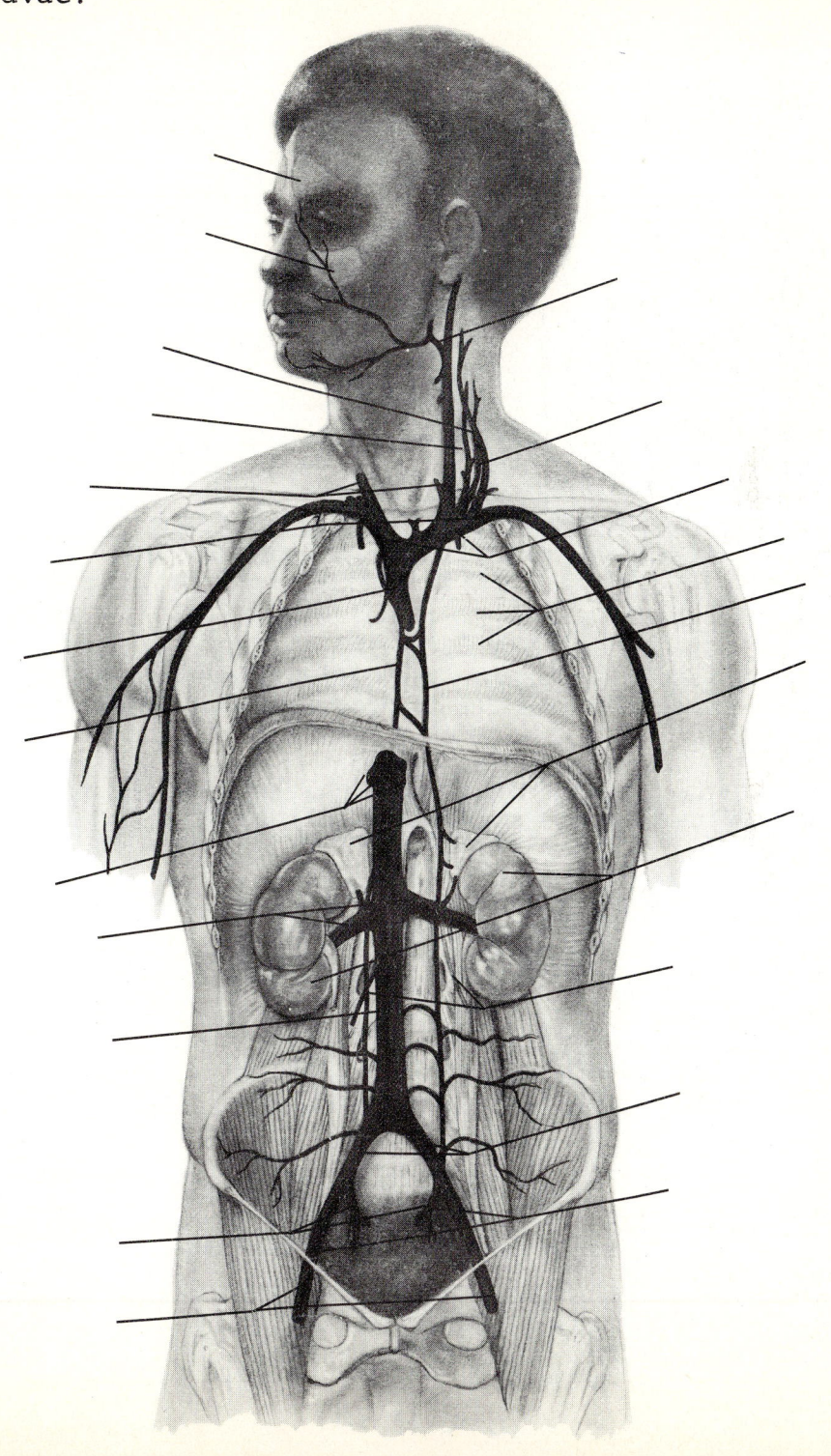

TEST YOURSELF

32. receives blood directly from the superior mesenteric v.___
33. anterior descending branch receives blood from the___
34. vertebral arteries arise from___

a. subclavian a.
b. internal carotid
c. left coronary a.
d. hepatic portal v.
e. hepatic sinusoids

> LO 14: Describe the heart including its pericardium, chambers, and valves; identify common valve deformities discussed and their consequences.

The heart wall consists of endocardium, myocardium, and epicardium. The pericardium is a tough connective tissue sac that surrounds the heart. The heart consists of right and left atria and right and left ventricles. Each atrium is separated from its ventricle by a valve, the tricuspid valve on the right and the mitral valve on the left. The aortic valve separates left ventricle from aorta and the pulmonic valve separates right ventricle and pulmonary trunk.

(1) Which valves are known as semilunar valves? _____

(2) Mitral stenosis is a narrowing of the passageway of the _____ _____

(3) Label the structures of the heart::
    (See figure on following page)

(4) What are chordae tendineae? _____
(5) What are papillary muscles? _____

TEST YOURSELF

35. Bulk of heart wall consists of___
36. Between left atrium and left ventricle___
37. Sac of fibrous connective tissue surrounding heart___

a. pericardium
b. myocardium
c. mitral valve
d. semilunar valve
e. tricuspid

> LO 15: Describe the structure and function of the heart's conduction system.

244

245

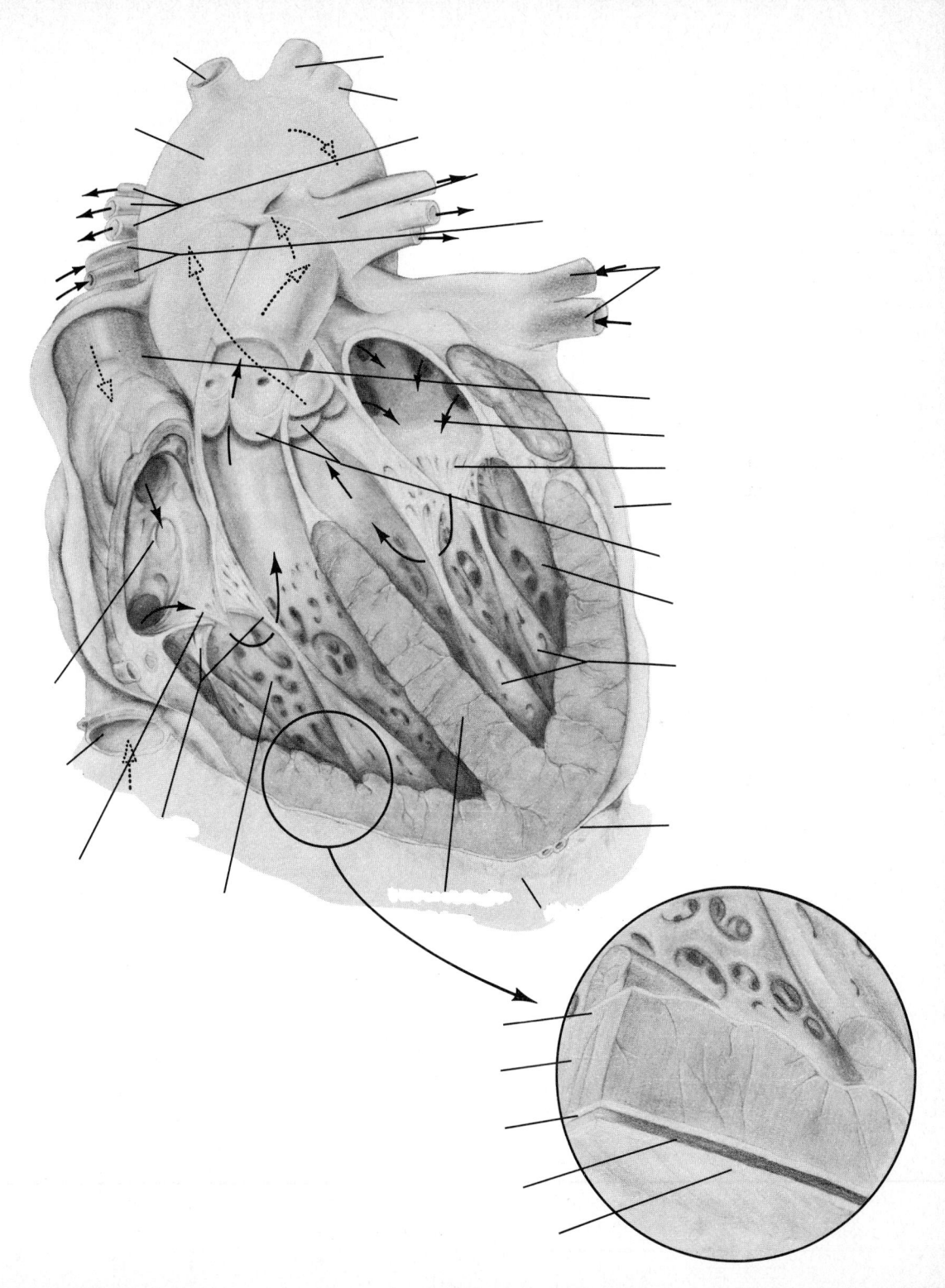

Each heartbeat begins in the SA node. The action potential spreads through the atria causing atrial contraction. One group of fibers conducts the action potential to the AV node; then it spreads through the Purkinje fibers and finally to the ordinary muscle of the ventricles.

(1) What does SA stand for?_____
(2) What does AV stand for?_____
(3) Which structure is known as the pacemaker of the heart?_____ Why?_____
(4) Give the location of:
   (a) the SA node_____
   (b) the AV node_____
(5) Describe the bundles formed by the Purkinje fibers:_____
_____

TEST YOURSELF
Match:

38. pacemaker   a. AV bundle
39. bundle of His   b. AV node
                              c. SA node

> *LO 16: Compare and contrast cardiac muscle with skeletal muscle.*

Like skeletal muscle, cardiac muscle is striated, has dark Z lines, and has myofibrils that contain actin and myosin and slide over one another during contraction. Unlike skeletal muscle, cardiac muscle is capable of contracting without neural stimulation; its fibers are branched and separated at their ends by intercalated discs which make cardiac muscle a functional syncitium.

Fill in the chart on the following page (246) by indicating <u>yes</u> where appropriate.

TEST YOURSELF
Match:
40. intercalated discs   a. Skeletal muscle only
41. branched fibers   b. Cardiac muscle only
42. contain myofibrils of   c. Both skeletal and
    actin and myosin   cardiac muscle

	Cardiac Muscle	Skeletal Muscle
Striated?		
Myofibrils?		
Branched fibers?		
Intercalated discs?		
Functional syncitium?		
Can contract without neural stimulation?		

LO 17: *List and briefly describe the events of the cardiac cycle, and correlate normal heart sounds with events of this cycle.*

Each cardiac cycle begins with the generation of an action potential in the SA node which causes atrial systole. As the atria contract blood is forced into the ventricles. Ventricular systole forces blood through the semilunar valves into the systemic and pulmonary circulations. At the same time the atria are in diastole and fill with blood. As the ventricles enter diastole the aortic and pulmonic valves close and the tricuspid and mitral valves open once again.

(1) When listening to the heart through a stethoscope the lub sound is heard when the_____and _____valves close.

(2) The dup sound is heard when the_____and_____ valves snap shut.

TEST YOURSELF

43. Blood is forced into the systemic and pulmonary circulations a. during ventricular diastole b. during ventricular systole c. during atrial systole d. at the time the lub sound is heard e. at the time the dup sound is heard

> LO 18: Correlate the principal waves of a
> normal ECG with the events of the cardiac
> cycle, and describe common arrhythmias
> and other disorders that can be diagnosed
> with the help of the ECG.

An electrocardiogram begins with a P wave which represents the spread of an impulse over the atria just prior to atrial contraction. Then the QRS complex indicates that the impulse is spreading over the ventricles just before ventricular contraction. As the ventricles recover, currents generated cause a T wave.

(1) Draw and label a normal ECG wave:

(2) Briefly describe:
    (a) tachycardia_____
    (b) bradycardia_____
    (c) flutter_____
    (d) fibrillation_____
    (e) heart block_____

(3) What can be done to help patients with severe heart-block syndromes?_____

TEST YOURSELF

44. A QRS complex on an ECG represents   a. ventricular fibrillation   b. first degree heart block   c. tachycardia   d. impulse spread over atria   e. impulse spread over ventricles

> LO 19: Describe the self-regulation and neural
> control of the heartbeat.

The heartbeat depends upon venous return, a normal balance of ions, and upon sympathetic and parasympathetic nerves which speed and slow its rate. Baroreceptors in the carotid sinuses and elsewhere are sensitive to changes in blood pressure and send messages to the brain providing feedback regarding needs for a slower or more rapid heartbeat.

(1) Define venous return:_____
(2) List three ions that may affect heart function:_____
(3) If the heart can beat on its own what purpose is served by its sympathetic and parasympathetic innervation?_____
(4) What are baroreceptors?_____
(5) How do they work?_____

TEST YOURSELF

45. When blood pressure <u>increases</u>, a. baroreceptors send impulses to cardiac centers in the medulla b. parasympathetic nerves are stimulated c. heart rate slows d. answers (a),(b) and (c) are correct e. answers (a) and (c) only are correct

> *LO 20: Define cardiac output and relate it to Starling's Law of the Heart.*

Cardiac output is the amount of blood pumped by one ventricle in one minute. According to Starling's Law, the more blood that is returned to the heart by the veins, the greater the volume of blood that will be pumped during the next systole.

(1) Explain the physiologic basis of Starling's Law

(2) Define stroke volume:_____
(3) What is heart failure?_____

TEST YOURSELF

46. The cardiac output is the volume of blood pumped by a. one ventricle during one beat b. both ventricles during one beat c. the heart of a trained athlete during one systole d. one ventricle during one minute e. the heart during a 24 hour period

> *LO 21: Identify factors that affect blood flow.*

Among the factors that affect blood flow are cardiac output, viscosity, elasticity of blood vessels, and diameter of blood vessels.

I. Analyze the equation
$$\text{Pressure} = \text{flow} \times \text{resistance}$$

(1) What constitutes flow?_____
(2) Resistance is presented mainly by the_____
(3) Blood flows fastest in the_____. Why?
    _____
(4) Why do veins offer little resistance to blood flow?
    _____
(5) Resistance increases in a blood vessel as the radius_____
            (increases, decreases)
(6) When one stands still for a long period of time blood pools in the_____
(7) This causes pressure to increase in the_____.
    Plasma moves out into the_____.
    Arterial blood pressure (rises, falls) and fainting may occur.

TEST YOURSELF

47. Flow is greatest when a. pressure is greatest
    b. pressure is lowest c. resistance is greatest
    d. resistance is lowest e. answers (a) and (d) are correct

> LO 22: Describe the homeostatic mechanisms that regulate blood pressure.

Blood pressure is regulated by neural and normonal mechanisms. The arterial baroreceptors provide feedback to the cardiac centers in the medulla, resulting in parasympathetic stimulation of the heart and ultimately decreased blood pressure.

(1) Baroreceptors also send impulses to the vasomotor center in the medulla. Explain how this acts to regulate blood pressure
(2) Summarize the steps by which the angiotensins work to regulate blood pressure:_____
    _____
(3) How does aldosterone affect blood pressure?_____
    _____

TEST YOURSELF

48. Angiotensin I is produced when a. stimulated by aldosterone b. the kidney is stimulated by the baroreceptors c. the kidneys release renin d. blood pressure in the kidneys increases e. three of the preceding answers are correct

> LO 23: *Give the physiologic basis of arterial pulse, and tell how pulse is measured.*

Pulse is a consequence of the elastic expansion and recoil of the arteries as they fill with blood; it is measured with sphygmomanometer and stethoscope.

(1) What happens to the aorta during ventricular systole? _____

(2) Then, what happens during diastole? _____

TEST YOURSELF

49. Jane's blood pressure was measured at 120/80. The 120 indicates a. that Jane has high blood pressure b. that Jane has low blood pressure c. the diastolic pressure d. the systolic pressure e. that she is anemic

> LO 24: *Describe the progressive pathologic changes of atherosclerosis, give its risk factors and common complications, and tell how it is treated.*

In atherostlerosis plaques consisting of arterial smooth muscle, cholesterol, and calcium develop within the intima of arteries, eventually impeding the circulation of blood. Risk factors of atherosclerosis include cigarette smoking, hypertension, and high blood-cholesterol levels.

(1) List 3 other risk factors: _____
_____

(2) Briefly describe aortic-coronary bypass surgery: _____
_____

(3) Give the physiologic basis for common complications of atherosclerosis:_____
_____
_____

TEST YOURSELF

50. Which of the following are risk factors for atherosclerosis: a. cigarette smoking   b. low serum cholesterol level   c. low blood pressure   d. high levels of estrogen   e. all of the preceding answers are correct

> LO 25: Describe other cardiovascular diseases discussed, including angina, hypertension, coronary occlusion, myocardial infarction, thrombosis, and embolism.

(1) Briefly describe each of the following:
    (a) Hypertension_____
    (b) Angina_____
    (c) Coronary occlusion_____
    (d) Mycardial infarction_____
    (e) Thrombosis_____
    (f) Embolism_____

TEST YOURSELF
Match:

51. death of a portion of cardiac muscle due to ischemia
52. pain due to exertion in person with ischemic heart disease
53. blockage of blood vessel due to intravascular clot that has moved from its point of origin

a. embolism
b. thrombosis
c. angina
d. myocardial infarction
e. hypertension

ANSWERS TO TEST YOURSELF QUESTIONS

1.c	2.d	3.a	4.c	5.d	6.b	7.a	8.c	9.e
10.b	11.c	12.a	13.a	14.c	15.e	16.d	17.e	
18.d	19.c	20.a	21.a	22.d	23.c	24.d	25.e	
26.c	27.b	28.a	29.d	30.d	31.c	32.d	33.c	
34.a	35.b	36.c	37.a	38.c	39.a	40.b	41.b	
42.c	43.b	44.e	45.d	46.d	47.e	48.c	49.d	
50.a	51.d	52.c	53.a					

POST TEST

1. (LO 1) Which of the following are functions of the cardiovascular system: a. transports nutrients b. transports oxygen c. produces aldosterone d. answers (a),(b) and (c) are correct e. answers (a) and (b) only are correct

2. (LO 2) Albumins and globulins are a. carbohydrates found in red blood cells b. lipids found in platelet cell membranes c. proteins found in the plasma d. sugars found in platelet mitochondria e. fatty acids in the plasma

3. (LO 3) Red blood cells are proliferated from a. stem cells in the bone marrow b. reticulocytes in the liver c. monocytes in the spleen d. methemoglobin in the kidney e. rennin in the kidneys

4. (LO 4) The function of REF (renal erythropoietic factor) is to stimulate a. hypoxia b. production of erythropoietin c. stem cells in the bone marrow d. white blood cell production e. plasma protein production

5. (LO 5,7) Mary has been feeling tired and easily fatigued. Her physician finds a low hematocrit and red cells that are hypochromic and microcytic. After questioning her about her diet he determines that she has a. acute leukemia b. chromic leukemia c. iron deficiency anemia d. macrocytic anemia e. sickle cell anemia

(LO 6-8 refer to questions 6-8)
6. megakaryocytes give rise to them
7. agranular leukocytes
8. adept at ingesting bacteria

a. platelets
b. basophils
c. neutrophils
d. monocytes
e. all of the above

9. (LO 9) In the formation of a clot the conversion of fibrinogen to fibrin is catalyzed by a. factor XII b. Hageman factor c. prothrombin d. vitamin K e. thrombin

10. (LO 10) Through a microscope you view a large blood vessel with a great deal of elastic connective tissue in the tunica media. You determine that the vessel is a   a. capillary   b. vein   c. sinusoid   d. artery   e. venule

(LO 11,12,13 refer to questions 11-14)

11. deliver(s) blood to aorta
12. deliver(s) blood to right atrium
13. deliver(s) blood to hepatic portal vein
14. extension(s) of aorta

a. common iliac arteries
b. vertebral arteries
c. coronary sinus
d. superior mesenteric veins
e. left ventricle

15. (LO 14) The cusps of the AV valves are anchored to the papillary muscles by the   a. intercalated discs   b. fossa ovalis   c. epicardium   d. chordae tendineae   e. pericardium

16. (LO 16) How does cardiac muscle differ from skeletal muscle   a. lacks striations   b. has branched fibers   c. lacks intercalated discs   d. lacks actin and myosin   e. has Z lines

17. (LO 17) Blood fills the ventricles to about 70 percent of their capacity   a. before atrial contraction occurs   b. after atrial contraction   c. during ventricular systole   d. just prior to the time the second sound (dup) may be heard with a stethoscope   e. two of the preceding answers are correct

18. (LO 15,18) Spread of an impulse over the atria is reflected on an ECG by   a. QRS complex   b. P wave   c. T wave   d. R wave   e. tachycardia wave

19. (LO 19) The heart rate may be <u>increased</u> by   a. decreased venous return   b. decrease in blood pressure   c. a greater number of impulses from the baroreceptors   d. release of acetylcholine   e. all of the preceding answers are correct

20. (LO 20) Normally cardiac output depends mainly upon   a. signals from sympathetic nerves   b. venous return   c. viscosity of blood   d. impulses entering

the Bundle of His from the atria muscle fibers   e. the number of calcium ions present in the coronary sinus

21. (LO 21) Blood flow is fastest through   a. arteries   b. arterioles   c. capillaries   d. sinusoids   e. venules

22. (LO 22) You do not faint when you rise from a lying position because   a. baroreceptors signal sympathetic nerves to blood vessels   b. vasoconstriction occurs   c. blood pressure is increased to compensate   d. answers (a),(b) and (c) are correct   e. none of the preceding answers is correct

23. (LO 23) The aorta expands   a. when it fills with blood   b. during ventricular systole   c. when one has low blood pressure   d. when signalled by baroreceptors   e. answers (a) and (b) only are correct

(LO 24,25 refer to questions 24-28)

24. may occur as a complication of atherosclerosis
25. often associated with angina
26. weakened wall of artery or heart that may balloon outward
27. intravascular clot
28. death of a portion of heart muscle

a. thrombosis
b. ischemic heart disease
c. myocardial infarction
d. aneurysm
e. all of the above

ANSWERS TO POST TEST QUESTIONS

1.e   2.c   3.a   4.b   5.c   6.a   7.d   8.c   9.e
10.d   11.e   12.c   13.d   14.a   15.d   16.b   17.a
18.b   19.b   20.b   21.a   22.d   23.e   24.e   25.b
26.d   27.a   28.c

14 THE LYMPHATIC SYSTEM

PRETEST

1. List 3 functions of the lymphatic system_____
   _____ _____
2. The cisterna chyli empties lymph into the_____
3. Lymph from the upper right quadrant of the body is returned to the right subclavian vein via the right _____
4. Two main functions of lymph nodes are_____ and_____
5. One way that lymph nodes differ from lymph nodules is that they are surrounded by a capsule of _____ _____
6. The largest organ of the lymphatic system is the _____
7. Some plasma leaves the circulation to form_____ _____; it leaves mainly because of _____
8. Excessive accumulation of interstitial fluid is called_____
9. Immunity depends upon the fact that each person is _____ unique
10. _____ is a protein released by cells in response to viral invasion
11. In cellular immunity the_____lymphocytes are the principal cells
12. In humoral immunity B cells produce specific _____
13. Immunologic competence is thought to be conferred upon T lymphocytes by the_____
14. The portion of an antigen that can be recognized by an antibody is called its_____

256

15. Cellular immunity is directed mainly at fungi and _____
16. In humoral immunity selected lymphocytes differentiate into _____ _____ which secrete large amounts of _____
17. Antibodies often do their work through the _____ system
18. In opsonization complement coats the surfaces of the _____, a process which enhances _____
19. _____ immunity is used to boost one's defenses temporarily against a particular disease
20. Vaccination is a form of_____ immunity
active, passive
21. IgE antibody mediates common_____ reactions
22. In_____ an allergen-reagin reaction occurs in the bronchioles of the lungs
23. In desensitization therapy patients are injected with small amounts of the _____
24. A person with type A blood has_____ antigens on the surfaces of his red blood cells and _____ antibodies in his plasma
25. The immune response aimed at transplanted tissues is known as_____ _____ and is launched by_____

ANSWERS: 1. returns excess interstitial fluid to the blood, defends body against disease organisms, absorption of lipids, (LO 1)   2. thoracic duct (LO 2)
3. lymphatic duct (LO 2)   4. filter lymph, produce lymphocytes (LO 3)   5. connective tissue (LO 4)
6. spleen (LO 5)   7. interstitial fluid, hydrostatic pressure (LO 6)   8. edema (LO 7)   9. biochemically (LO 8)   10. interferon (LO 9)   11. T (LO 10)
12. antibodies (LO 10)   13. thymus (LO 11)   14. antigenic determinant (or epitope) LO 12)   15. viruses (or parasites) (LO 13)   16. plasma cells, antibody (LO 14)
17. complement (LO 15)   18. pathogens, phagocytosis (LO 15(19. Passive (LO 16)   20. active (LO 17)
21. allergic (LO 18)   22. asthma (LO 18)   23. antigen to which they are allergic (LO 19)   24. A,B (LO 20)
25. graft rejection, T lymphocytes (LO 21)

LO 1: List and briefly describe the three major functions of the lymphatic system.

Three functions of the lymphatic system are:
1._____

2._____

3._____

TEST YOURSELF

1. Which of the following is NOT a function of the lymphatic system: a. responsible for immunity b. lipid absorption c. production of platelets d. returns excess interstitial fluid to blood

LO 2: Identify the main structures of the lymphatic system and describe the overall plan of lymphatic circulation.

Main structures of the lymphatic system are (1) lymphatic vessels, (2) lymph, (3) lymph nodes and nodules, (4) thymus and spleen.

(1) List two types of vessels in the lymphatic system
    (a)_____

    (b)_____

(2) Trace a drop of lymph from a lymph capillary to the left subclavian vein.
    lymph capillary ⟶

TEST YOURSELF

2. Which of the following is NOT a part of the lymphatic system a. lymphatics b. lymph capillaries c. lymph heart d. cisterna chyli e. lymph nodes

LO 3: Give two main functions of lymph nodes, describe their structure, and tell where the lymph nodes are located.

Lymph nodes are distributed along the main lymphatic routes especially in the axilla, groin, neck, thorax, and abdomen.

(1) Two functions of lymph nodes are: 1._____
    2._____
(2) Label the lymph node:

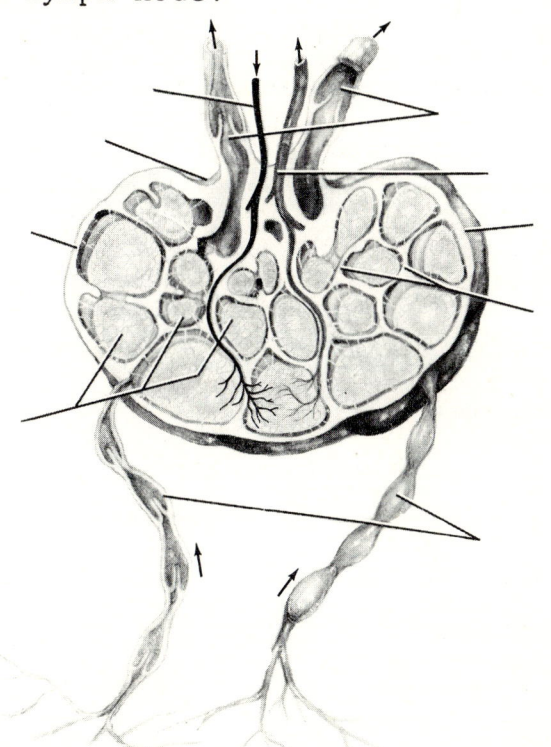

TEST YOURSELF

3. Lymph nodes are numerous in the a. axilla b. groin c. fingers d. answers (a),(b) and (c) are correct e. answers (a) and (b) only are correct

LO 4: Contrast lymph nodules with lymph nodes and give their function.

Lymph nodules are not surrounded by connective tissue capsules and do not have afferent lymphatic vessels.

They are scattered throughout connective tissue and are especially numerous in the walls of the intestine, upper respiratory passages, and urinary tract.

(1) What is the function of lymph nodules?_____

(2) Where are Peyer's patches located _____

TEST YOURSELF
Match

4. have connective tissue capsules
5. lack afferent lymphatic vessels
6. produce lymphocytes

a. lymph nodules
b. lymph nodes
c. both

> LO 5: Identify tonsils, spleen, and thymus as lymph organs, and give their functions.

Tonsils are masses of lymph tissue sometimes classified as lymph nodules. The spleen is the largest of the lymphatic organs. The thymus is located in the mid-upper portion of the chest cavity behind the sternum.

Give the functions of
(a) tonsils_____
(b) spleen_____
(c) thymus_____

TEST YOURSELF
Match

7. stores platelets, produces phagocytes, filters blood
8. decreases in size after puberty; located in chest cavity

a. tonsils
b. spleen
c. thymus

> LO 6: Describe the relationship between plasma, interstitial fluid, and lymph: contrast differences in composition and tell how each is formed.

Interstitial fluid forms when plasma is forced out of the capillaries as it enters from arterioles. Its composition is similar to plasma without cells and with little protein. Lymph is interstitial fluid that has *found its way into lymph vessels. As lymph circulates through lymph nodes it accumulates large numbers of lymphocytes.*

(1) Define and tell how each force affects flow of fluid between capillary and interstitial fluid:
   (a) hydrostatic pressure_____

   (b) colloid osmotic pressure_____

   (c) effective filtration pressure_____

(2) State Starling's law of the capillaries:_____

(3) How does construction of the lymphatic capillary wall facilitate formation of lymph?_____

(4) How is it that the lymph capillary is able to exert suction on the interstitial fluid?_____

TEST YOURSELF

9. The wall of a lymph capillary may best be described as a. an open-ended straw b. a tiny hose riddled with pores c. a blind tube with tiny one-way swinging doors that open inward d. a leaky tube lined with Kuppfer cells e. an-open-ended tube with a thick layer of smooth muscle in its wall

> *LO 7: Identify the mechanisms responsible for lymph flow: define edema and describe its cause.*

Mechanisms that contribute to lymph flow are the pumping action of the lymph vessels, the presence of valves, muscle contraction, arterial pulse, and other factors resulting in compression of body tissues.

(1) Define edema:_____
(2) Briefly explain how each of the following conditions may result in edema:
   (a) cardiac failure_____
   (b) radical mastectomy_____
   (c) filariasis_____

TEST YOURSELF

10. Edema may result from a. blockage of lymph vessels
    b. surgical removal of lymph vessels   c. ineffective
    pumping by the heart d. answers(a),(b) and (c) are
    correct e. answers (a) and (b) only are correct

>   LO 8: Identify biochemical individuality as
>   the basis of immunity.

Immunity depends upon the fact that each individual has his own specific array of macromolecules which are slightly different than everyone else's molecules. The body comes to recognize its own molecules and targets all others as foreign and therefore potentially destructive.

(1) Define immunity_____
(2) What are foreign macromolecules called?

_____

TEST YOURSELF

11. Cells responsible for immunity a. "know" their
    own body's macromolecules b. recognize the macro-
    molecules from other organisms as foreign c. launch
    immune attacks when stimulated by foreign macro-
    molecules d. answers (a),(b) and (c) are correct
    e. answers (a) and (b) only are correct

>   LO 9: Describe four nonspecific defense
>   mechanisms, including the reticuloendothelial
>   system, inflamatory reaction, nutritional
>   immunity and interferon.

Nonspecific defense mechanisms prevent entry of pathogens and attempt to destroy them quickly should they enter the body.

Briefly describe, and tell how each acts to defend the body against pathogens
    (a) the reticuloendothelial system_____

    (b) inflammation_____

(c) nutritional immunity_____

   (d) interferon_____

TEST YOURSELF
Match:

12. tissue macrophages   a. inflammation
13. blood concentration of   b. nutritional immunity
    iron lowered   c. skin
                                d. interferon
                                e. reticuloendothelial
                                   system

> LO 10: Define the terms: antigen and antibody;
> tell how one antigen differs from another and
> how antigens stimulate immune responses.

Each antigen has several distinct antigenic determinants (epitopes) upon its surface. These can be recognized by antibodies, or specific cell receptors. When an antigen enters the body it stimulates the type of lymphocytes specifically programmed to react to its structure.

(1) a. define antigen_____
    b. define antibody_____
2. what type of molecular subunits determine the
   antigenic nature of a protein-type antigen?
   _____
3. what is a hapten?_____

TEST YOURSELF

14. Small molecules that combine with carrier proteins
    to become antigens are   a. haptens b. epitopes
    c. antigenic determinants d. thymosin producers
    e. two of the preceding answers are correct

> LO 11: Describe how antibodies act directly
> upon pathogens and how they act by way of
> the complement system.

Antibodies can combine with antigens to form an antigen-antibody complex which is then phagocytized by

macrophages. Antibodies often activate the complement
system which helps attack pathogens in several ways
inclduing by enhancing inflammation and by opsonization.

(1) Briefly describe the role of each class of antibody:
    (a) IgG_____
    (b) IgM_____
    (c) IgA_____
    (d) IgE_____

(2) What is the complement system?_____

(3) What happens when anaantibody fixes complement?_____
_____

(4) How does complement increase inflammation?_____

(5) Describe opsonization_____
_____

TEST YOURSELF

15. The complement system a. consists of several pro-
    teins found in plasma b. is activated when IgG or
    IgM has combined with an antigen c. is highly
    specific, i.e. each component reacts with only one
    type of antigen d. answers (a),(b) and (c) are
    correct e. answers (a) and (b) only are correct

> LO 12: Describe the mechanism of cell mediated
> immunity, including the development of mem-
> ory cells.

When sensitized by a specific antigen, T lymphocytes
increase in size, multiply to produce a clone of identical cells, then migrate to the site of invasion.
There they attempt to destroy the invader. Some sensitized T cells remain in the lymph nodes and live on
as memory cells.

(1) How many types of T lymphocytes are present in the
    body?_____
(2) What role do macrophages play in sensitizing T
    cells?_____
_____

(3) How do T lymphocytes destroy pathogens?

(4) What are lymphokines?
(5) What is the theory of immunosurveillance?

TEST YOURSELF

16. T lymphocytes   a. are responsible for mediating humoral immunity   b. are processed in the thymus   c. can be traced originally to stem cells in the spleen   d. answers (a),(b) and (c) are correct   e. answers (b) and (c) only are correct

17. A virus with antigen X enters the body. Which of the following events would occur   a. all of the lymphocytes begin to produce clones which will attack the pathogen   b. only the type of lymphocytes programmed to attack antigen X are activated   c. lymphokines are released from the lymph glands and reach the site of invasion via the lymph   d. none of the preceding answers are correct   e. answers (b) and (c) only are correct

> *LO 13: Describe the mechanism of humoral immunity.*

When activated by a specific antigen, B lymphocytes multiply producing a large clone. These cells differentiate to become plasma cells which secrete large amounts of antibody. Some B cells do not fully differentiate and become memory cells.

(1) How many different types of B cells exist in the body?
(2) In what ways do B cells respond differently than T cells?

(3) What are helper T cells?
(4) What role in immunity is played by memory cells?

TEST YOURSELF

18. When B cells are activated they a. rush to the site of invasion and shower the pathogen with antibodies b. multiply and produce plasma cells which release antibodies into the circulation c. form several types of helper cells which produce complement d. develop into macrophages which phagocytize the invaders e. three of the preceding answers are correct

    LO 14: Describe the role of the thymus in immune mechanisms.

The thymus is believed to confer immunologic competence upon T lymphocytes and also secretes thymosin, a hormone thought to stimulate T lymphocytes so that they become immunologically active.

(1) When does the thymus "instruct" T lymphocytes?

(2) What happens when the thymus is removed from an experimental animal at birth (before the thymus has conferred immunologic competence upon the T cells)?

(3) Children suffering from DiGeorge's syndrome suffer from a deficiency in humoral as well as cellular immunity. Why?

TEST YOURSELF

19. The thymus is thought to a. instruct T lymphocytes just before birth and during the first few months of postnatal life b. stimulate the kidneys to produce thymosin c. "teach" B cells how to produce antibodies d. answers (a),(b) and (c) are correct e. answers (a) and (b) only are correct

    LO 15: Distinguish between active and passive immunity, giving examples of each and explaining why passive immunity is temporary.

In active immunity the body is exposed to an antigen and actively launches an immune response so that memory cells are produced. When a person is injected with a vaccine his body goes through all of the steps involved in establishing immunity just as if he had actually been exposed to the pathogen in a natural way. Although artificially induced, vaccination is a form of active immunity. In passive immunity antibodies are borrowed from another person or animal which has produced them.

(1) Distinguish between natural and artificial active immunity: _____

(2) How are vaccines prepared? _____

(3) Are memory cells produced in passive immunity? _____

When is passive immunity used clinically? _____

TEST YOURSELF

20. A pregnant woman is exposed to German measles. Her physician injects her with gamma globulin. Which of the following is (are) true a. she has received passive immunity b. her immunity to German measles will last for many years c. she has received artificial, active immunity d. she has received live, attenuated German measles virus e. answers (b) and (c) are correct

21. Vaccines produced from attenuated viruses a. confer passive immunity b. often give the patient the actual disease c. stimulate the mechanisms involved in establishing active immunity d. are no longer marketed because they are dangerous e. answers (b) and (d) are correct

> *LO 16: Give the immunologic basis of allergy, and briefly describe the following allergic reactions: asthma, hayfever, hives, and systemic anaphylaxis.*

In a common type of allergic reaction an allergen stimulates production of an IgE. The immunoglobulin (antibody) then combines with receptors in the surfaces of mast cells (tissue basophils). The mast cell releases histamine and other chemicals which cause vasodilatation and increased capillary permeability.

(1) Define:
   (a) reagin_____
   (b) allergen_____
   (c) hypersensitivity_____
(2) Describe the physiologic basis for
   (a) asthma_____
   (b) hayfever_____
   (c) hives_____
   (d) systemic anaphylaxis_____

TEST YOURSELF
Match:
22. may result in circulatory shock; widespread reaction
23. slow-reacting substance released from mast cells in bronchioles of lungs

a. asthma
b. hives
c. hayfever
d. systemic anaphylaxis
e. IgA reaction

>LO 17: Give the physiologic basis for using antihistamines and desensitization therapy in treating allergic disorders.

Antihistamines block the effect of histamines. In desensitization therapy, the patient is injected with the very antigen to which he or she is allergic.

(1) How do antihistamines block the effects of histamines?_____

(2) What is the rationale of injecting an allergic patient with the antigen to which she is allergic? Explain immunologically how this can help.
_____
_____
_____

TEST YOURSELF

24. The purpose of desensitization therapy is to a. stimulate production of IgG antibodies against the allergen b. suppress production of IgE by getting the body used to the allergen c. block the effects of histamine d. answers (a),(b) and (c) are correct e. answers (b) and (c) only are correct

> LO 18: Describe the ABO and Rh blood-type systems, and t ll why blood types must be carefully matched in transfusion therapy.

(1) Give the antigen and antibody associated with each blood type:

Blood type	Antigen	Antibody
A	_____	_____
B	_____	_____
O	_____	_____
AB	_____	_____
Rh	_____	_____

(2) What is the cause of erythroblastosis fetalis? Under what circumstances might it occur?

(3) How can the disorder be guarded against?

TEST YOURSELF
Match:
25. has A and B antigen
26. has no ABO antibodies
27. has antigen A and antibody B

a. Type A
b. Type B
c. Type AB
d. Type O

28. Erythroblastosis fetalis may occur when an Rh____ man impregnates an Rh____ woman: a. -, + b. -,- c. + or -,-   c. + or -, d. +, -   e. +,+

LO 19: Describe the problem of graft rejection, and tell how physicians minimize its effects in transplant patients.

When an organ is transplanted from one person into another, the recipient's body regards the graft as foreign and launches an immune response against its antigens.

(1) What is the HL-A group of antigens?_____
(2) How do physicians counter the problems of graft rejections when they occur?_____

(3) What is an immunologically privileged site? Give an example:

TEST YOURSELF

29. Histocompatability antigens a. are the ones that affect transplants b. are identical in identical twins c. include the HL-A group d. answers (a), (b) and (c) are correct e. answers (a) and (c) only are correct

30. Physicians minimize the effects of graft rejections by using a. radiation that destroys lymphocytes b. drugs that destroy lymphocytes c. serum prepared from HL-A group antigens d. answers (a), (b) and (c) are correct e. answers (a) and (b) only are correct

ANSWERS TO TEST YOURSELF QUESTIONS

1.c	2.c	3.e	4.b	5.a	6.c	7.b	8.c	9.c
10.d	11.d	12.e	13.b	14.a	15.e	16.b	17.b
18.b	19.a	20.a	21.c	22.d	23.a	24.a	25.c
26.c	27.a	28.d	29.d	30.e

POST TEST

1. (LO 1) The lymphatic system a. is responsible for immunity b. returns excess interstitial fluid to the circulation c. absorbs lipids from the GI tract d. answers (a),(b) and (c) are correct e. answers (a) and (c) only are correct

2. (LO 2) The thoracic duct delivers lymph into the
   a. cisterna chyli   b. right lymphatic duct
   c. spleen   d. lymph capillaries of the chest
   e. left subclavian vein

   (LO 3,4,5) refer to questions 3-6
3. distributed along the main lymph-   a. lymph node
   atic routes (e.g. in axilla)   b. lymph nodule
4. found in connective tissue; small   c. spleen
   structure not surrounded by   d. thymus
   capsule   e. none of the
5. large lymph organ that produces   above
   phagocytes and stores platelets
6. filters lymph; small lima bean shaped structure

7. (LO 6) The colloid osmotic pressure of the blood
   a. forces plasma out of the capillary   b. acts to
   retain plasma within the capillary   c. results
   from plasma proteins   d. is responsible for drawing
   protein back into the capillary at the venous end
   of a capillary network   e. answers (b) and (c)
   are correct

8. (LO 6) The net force that determines the actual
   movement of fluid between blood and interstitial
   fluid is the   a. hydrostatic pressure of the blood
   b. hydrostatic pressure of the interstitial fluid
   c. effective filtration pressure   d. osmotic pressure of the blood   e. answers (a) and (b) are
   correct

9. (LO 7) Excessive accumulation of interstitial fluid
   is termed   a. filtration   b. filariasis   c. lymph-
   otosis   d. derangement of Starling's Law   e. edema

10. (LO 7) Which of the following contribute to lymph
    flow   a. muscle contraction   b. valves within the
    lymph vessels   c. pumping action of the lymph ves-
    sel   d. answers (a),(b) and (c) are correct
    e. none of the preceding answers is correct

11. (LO 8: Immunity depends upon the fact that   a. each person is biochemically unique   b. viruses are characterized by specific toxins that the immune system recognizes as poisonous   c. all pathogens are coated with interferon that stimulates immune responses   d. pathogens stimulate eosinophils   e. two of the preceding answers are correct

12. (LO 9) The reticuloendothelial system   a. includes tissue macrophages and lymphocytes   b. is characterized by highly mobile cells   c. is a special subsystem of the lymphatic system   d. answers (a),(b) and (c) are correct   e. (a) and (c) only are correct

13. (LO 9) Interferon   a. inhibits multiplication of viruses   b. attracts neutrophils during inflammatory responses   c. is the form in which iron is stored in the liver during nutritional immunity   d. is responsible for diapedesis of macrophages   e. two of the preceding answers are correct

14. (LO 10) An antigenic determinant   a. can be recognized by a specific antibody   b. would be found upon the surface of an antibody   c. may be any of five varieties   d. is usually composed of fatty acids   e. all of the preceding answers are correct

15. (LO 11) Antibody B has fixed complement. This means that   a. the antigen has combined with an antigen upon the pathogen permitting the complement to bind to the pathogen surface   b. complement is able to act upon the pathogen   c. complement B has chemically combined with antibody B   d. IgE has combined with the complement   e. answers (a) and (b) are correct

(LO 12,13,14) refer to questions 16-20
16. responsible for mediating cellular immunity
17. differentiate into plasma cells
18. present antigens to the appropriate cells of the immune system
19. processed in the thymus gland
20. production of antibodies

a. B lymphocytes
b. T lymphocytes
c. both T and B lymphocytes
d. neutrophils
e. macrophages

21. (LO 15) Amy is vaccinated against rubella. She has received a form of   a. passive immunity   b. natural, active immunity   c. gamma globulin   d. artificial, active immunity   e. temporary immunity

22. (LO 16) A person who produces reagins   a. is hypersensitive to an allergen   b. is allergic   e. lacks IgE   d. answers (a) and (b) only are correct   e. (a),(b) and (c) are correct

23. (LO 17) Ken has received desensitization therapy against ragweed pollen   a. he has been injected with small amounts of the pollen antigen   b. now when he comes in contact with ragweed pollen he will produce IgG antibodies against the antigens   c. IgE will no longer be able to combine with the allergen   d. answers (a),(b) and (c) are correct   e. answers (a) and (c) only are correct

24. (LO 18) Bob has blood type A. In transfusion therapy he is accidentally given blood from a Type B donor. What is likely to happen?   a. nothing because these two blood types are compatible   b. anti-B in Bob's blood may combine with the B antigens on the donated red cells causing them to agglutinate and hemolyze   c. anti-A in the donated blood will destroy Bob's platelets   d. Anti-A in the donated blood will attack Bob's white blood cells   e. none of the preceding ansers is correct

25. (LO 19) Tom has received a kidney transplant and subsequently exhibited symptoms of graft rejection. His physicians treated him with X-ray. As a result   a. his T lymphocytes have been selectively destroyed   b. both T and B kidney lymphocytes have been weakened   c. he is vulnerable to pneumonia and other infections   d. his HL-A antigens have been modified so that he will accept the graft

ANSWERS TO POST TEST QUESTIONS
1.d   2.e   3.a   4.b   5.c   6.a   7.e   8.c   9.e
10.d   11.a   12.e   13.a   14.a   15.e   16.b   17.a
18.e   19.b   20.a   21.d   22.d   23.d   24.b   25.c

15 THE RESPIRATORY SYSTEM

PRETEST

1. After passing through the larynx, inhaled air enters the _trachea_
2. The nasal conchae _moisten_ the inhaled air
3. In the alveoli the air gains _carbon dioxide_ and loses _oxygen_, but the capillary blood gains _oxygen_ and loses _carbon dioxide_
4. The lungs charge the _blood_ with oxygen needed for cellular respiration, and also remove _carbon dioxide_ from it, neither of which could be accomplished without them
5. The nasal cavity is entered by way of the two external _nares_. Within, it is lined with a sticky sheet of _mucus_ which normally drains into the _pharynx_ and is then _swallowed_
6. The sinus cavities are located in center of the _skull bones_. They communicate via meati with the _nasal cavities sinusitis_
7. The pharynx is divided into three regions, the _oropharynx_, the _nasopharynx_ and the _laryngopharynx_. The soft palate divides the _oropharynx_ from the _nasopharynx_ (Answer this questions by listing anterior structures first).
8. The larynx consists of several cartilages, articulated by _synovial_ joints, and contains the _vocal cords_ responsible for voice production
9. You try to laugh as you swallow water and some water slips through the glottis. Your larynx immediately _closes_
10. The tone of the voice is controlled by vertical movement of the _larynx_ and by varying the tension of the _vocal cords_

trachea and into the bronchi, then into the alveoli. In expiration this route is reversed. The air is humidfied and filtered by the nasal cavity and there and in the chest tubes it is cleaned of most impurities by the cilia-mucus escalator system.

(1) In the space below, draw a diagram of the respiratory pathway, and label each structure.

(2) Humidification is mostly accomplished by the _____
(3) Gas exchange takes place in the _____
(4) Most of the respiratory tree is lined with sticky _____ that is moved by _____ epithelium.
(5) Give the function(s) for each structure listed in the table: (see following page)

TEST YOURSELF

1. after passing the trachea, an inhaled molecule of oxygen would enter   a. bronchus   b. a lymph node in the lung   c. the larynx   d. the pharynx   e. an alveolus

structure	function
(a) nose	
(b) pharynx	
(c) glottis	
(d) larynx	
(e) trachae	
(f) bronchi	
(g) bronchioles	
(h) air sacs (alveoli)	

TEST YOURSELF cont'd

2. If you inhale through your mouth instead of your nose, a. air passes into the esophagus instead of through the glottis b. air bypasses the pharynx but passes through the eustachian tube directly into the trachea c. air passes through the pharynx and glottis into the larynx d. air is shunted through the roof of the mouth into the nose so that it can be filtered e. less air can reach the respiratory system

3. The trachea is popularly known as the a. voice box b. windpipe c. air-sacs d. lung e. throat

4. Which of the following is (are) not found in the lungs? a. alveoli b. bronchioles c. trachea d. capillary networks e. two of the above

5. The nose functions to a. filter incoming air b. bring inhaled air to body temperature c. humidify incoming air d. answers (a),(b) and (c) are corre e. answers (a) and (b) are correct

>LO 2: Give three reasons why the respiratory system is essential for the life of the human body.

The respiratory system provides oxygen to the blood which in turn supplies it to the body cells. Also the respiratory system excretes the CO_2 carried to it by the blood from those cells. The body is far too large for these functions to be carried out by diffusion alone. Finally, the respiratory system is essential to the control of blood acid-base balance.

(1) The _____ of the human body renders direct access of the cells to air impractical.
(2) The blood, which transports _____ to the body cells and _____ from them, has no means of _____ exchange except the respiratory system.
(3) The respiratory system, by removing CO_2 from the blood at different rates, is able to contribute to the control of blood _____

TEST YOURSELF

6. The function of the respiratory system is a. to produce body energy b. to transport oxygen to the body cells and CO_2 away from them c. to provide for gas exchange between the air and blood d. to excrete carbon dioxide e. (c) and (d)

> LO 3: Describe the structure and give the main functions of the nasal cavity.
> LO 4: Describe the structure of the sinus cavities.
> LO 5: Describe sinusitis.

The double nasal cavity is entered by the external nares and opens into the pharynx by the internal nares. The conchae humidify the air and the internal layer of mucus traps microorganisms and particulate matter. The paranasal sinuses are cavities in some of the anterior skull bones which communicate with the nasal cavity by means of short meati. The sinuses have no known function but are vulnerable to sinusitis, an inflammatory condition in which the meati swell shut, interfering with their drainage.

(1) The _____ _____ divides the two nasal cavities from one another.

(2) The _____ are internal projections found by the turbinate bones which function to warm and to _____ the entering air.

(3) What function is served by the irregularity of the nasal cavity interior? _____

(4) What is the ultimate fate of nasal mucus and all that it may contain? _____
_____

TEST YOURSELF

7. The nasal cavities are divided from each other by the   a. conchae   b. septum   c. meatus   d. sinuses   e. turbinate bones

8. Which of the following is not a function of the conchae?   a. warms air   b. humidifies   c. induces turbulance   d. prevents entrance of coarse particles into the nasal cavities

>   LO 6: Describe the structure of the pharynx
>   (see Chapter 11).

The pharynx is composed of three regions: The anterior (1) oropharynx which is partitioned from the (2) nasopharynx by the soft palate and the (3) larngopharynx dorsel to the larynx. The pharynx contains the orifides of the fauces, the eustachian tubes, the posterior nares and the openings of the larynx and esophagus.

(1) Give the three regions of the pharynx
   1. _____
   2. _____
   3. _____

(2) Pharyngeal masses of lymphatic tissue are called _____. There are three pairs, roughly encircling the pharynx. These are the _____, the _____ and the _____ tonsils.

(3) Why does a sore throat often seem to stuff up the ears? _____

(4) The glottis, the _____ and the _____ work together to keep swallowed food or drink out of the trachea.

TEST YOURSELF

9. Which of these is a pharyngeal structure? a. external nares   b. turbinate bones   c. eustachian tubes   d. vocal cords   e. none of the preceding.

> LO 7: Describe the structure, functions, and common disorders of the larynx.

The larynx consists of several articulated cartilages united by synovial joints and attached to a complex set of muscles. The larynx contains the vocal cords, whose tension is regulated by intrinsic and extrinsic muscles. The vocal cords serve to close the glottis in swallowing, coughing and at certain other times, and in phonation. In phonation, they produce the voice by vibrating in the airstream.

(1) The pitch of the voice is regulated by
    1._____
    2._____
(2) In swallowing, the vocal cords_____
(3) The larynx has a skeleton of_____
(4) In the space below, fully describe the Heimlich maneuver

TEST YOURSELF

10. The vocal cords function in a. voice production
    b. defecation   c. swallowing   d. all of preceding
    e. (a) and (c) but not (b)

11. The most highly recommended way to rescue someone
    who is choking on food is by a. squeezing the ab-
    domen strongly   b. slapping him on the back
    c. offering a drink of water to restart the swallow-
    ing reflex   d. pulling the obstruction out with the
    finger or with a special instrument, widely avail-
    able in restaurants   e. giving mouth-to-mouth
    artificial respiration

> LO 8: Describe the structure and function
> of the trachea and bronchi with emphasis
> on the cilia-mucus escalator.

The trachea and bronchi are composed of very thin
connective tissue strengthened at intervals by
C-shaped rings of cartilage. Both are lined with a
mucus-secreting, ciliated epithelium, in which inhaled
particles are trapped.

(1) The trachea is a thin walled tube composed of
    _____ tissue and lined with _____
    epithelium.
(2) The trachea is strengthened at intervals by C-
    shaped _____. The _____ muscle
    runs the posterior length of the trachea, occupying
    all the gaps.
(3) The trachea branches into two similar _____.
    These in turn subdivide into yet smaller _____,
    which in their turn eventually divide to form the
    _____, which are devoid of cartilage and
    mucus glands.

TEST YOURSELF

12. Bronchi may be distinguished from bronchioles in
    that bronchi have _____ which bronchioles
    lack   a. connective tissue support   b. macrophages
    c. mucus glands   d. hyoid bones   e. conchae

LO 9: Describe the structure and function
of the lung with particular reference to:
  a. The bronchioles
  b. The terminal respiratory units
  c. The alveoli
  d. Gas transport and gas exchange

Smaller branches of the bronchi penetrate the substance of the lung and carry air within it. Near their ends, these lose their cartilage, eventually becoming respiratory bronchioles and finally alveoli. The alveoli are thin walled, elastic sacs infiltrated with capillaries. It is here that gases are exchanged between the atmosphere and the blood.

(1) The terminal respiratory units each consist of a respiratory_____plus a number of sac-like _____, surrounding a central_____.
(2) The alveoli would collapse from the effect of surface tension except for the presence of_____, which they secrete.
(3) In an alveolus,_____layers of cells generally lie between the air and the blood. Alveoli intercommunicate by the_____ _____
(4) The bronchial circulation is a branch of systemic/pulmonary circulation which supplies the bronchi and associated structure with blood.

TEST YOURSELF
(Some of these may have more than one correct answer)

13. Respiratory bronchioles
14. Atrium
15. Alveoli
16. Pores of Kohn
17. Bronchi

a. interconnect alveoli
b. possess mucus glands and cartilage
c. gas exchange takes place here
d. cleansed only by macrophage action

LO 10: a. Compare the composition of inhaled and exhaled air, describe the exchange of gases between alveoli and blood, and between blood and tissues. Describe how hemoglobin transports oxygen and CO_2 in the blood.

LO 10b. Summarize the principles of gas dynamics relevent to the alveolar and tissue exchange of gases.

In a mixture of gases the total pressure exerted is the sum of the partial pressures of the individual gas gases, so that each gas exerts a pressure proportional to its fractional content in the mixture. The partial pressure of each gas is also proportional to the total pressure of the mixture.

When a gas diffuses into a liquid, its final concentration in that liquid depends upon (1) its partial pressure in the gaseous state and (2) its solubility in the liquid. Thus at equilibrium a gas will have the same partial pressure in an aqueous liquid as in the adjacent air, but since gases are usually fairly insoluble in water, its proportional or per cent representation will be less, especially in the case of oxygen. Thus under ordinary conditions blood plasma alone cannot supply the body cells with enough oxygen to sustain life.

The hemoglobin of red cells progressively removes oxygen from its solution in the plasma of the lung capillaries even as more diffuses in from the atmosphere. Long before the blood has entered the pulmonary vein it therefore contains 20 vol % of oxygen, most of this in chemical combination with the hemoglobin inside the red cells.

The train of events is reversed in the tissue capillaries partly by reduction in partial pressure resulting from oxygen consmmption by the body cells and partly by the action of CO_2 produced by those cells. The CO_2 combines with hemoglobin, inducing a conformational change that liberates oxygen (The Bohr effect). Plasma and hemoglobin transport CO_2 to the lungs where, with the help of the enzyme carbonic anhgdrase, it leaves the blood.
Intra-red-cellular CO_2 forms H_2CO_3, which in turn dissociates to H^+ and HCO_3^-. The H^+ is absorbed by the hemoglobin so that only HCO_3^- remains. This bicarbonate diffuses out of the red cell into the plasma and is replaced by cl^- (cloride shift).

(1) If P = the total pressure of a mixture of gases, and P_1, P_2, P_3 and so on represent the partial pressure of the individual gases, write a formula in the space below that represents their relationship.

(2) The atmosphere in a spacecraft consists of pure oxygen at a pressure of 150 mmHg. Will the astronauts suffer from oxygen poisoning?_____
Why, or why not?_____

(3) The proportion of a water solution of a gas that is comprised by that gas depends upon (1)_____
(2)_____

(4) When oxygen combines with Hb, this has the effect of_____ it from solution so that more _____ dissolves in the plasma. If the Hb is as yet not all chemically oxygenated, this new oxygen will in turn combine with_____, and so on.

(5) The diffusion of two gases, i.e. O_2 and CO_2, is mutually influenced/unaffected. In consequence, since their proportional_____ and different in alveolar blood,_____ diffuses into the atmosphere.

(6) What is the Bohr effect?_____

(7) In the space blow, develop a <u>diagram</u> of the chloride shift

(8) To sum up: (a) Gas exchange between air sacs and blood and between blood and cells takes place by _____ (b) How is oxygen transported in the blood?_____
_____ (c) How is carbon dioxide transported?
_____ (d) There is less oxygen in exhaled than in inhaled air because_____
_____

285

(e) There is more carbon dioxide in exhaled than in inhaled air because _____

TEST YOURSELF

In the following questions, (nos. 18-23) indicate a movement <u>away</u> with the letter <u>a</u>, no movement with <u>b</u>, and movement <u>towards</u> with <u>c</u>.

18. Oxygen-alveolar blood
19. CO_2-alveolar blood
20. Oxygen-alveolar hemoglobin
21. CO_2-tissue capillary hemoglobin
22. Chloride-alveolar red cells
23. Hydrogen ions-tissue capillary homoglobin

24. Carbonic anhydrase <u>promotes</u> H_2CO_3 formation
25. Carbonic anhydrase <u>breaks</u> up H_2CO_3
26. Bohr effect liberates oxygen
27. Plasma absorbs 20 vol % of dissolved O_2

a. Alveolar capillaries
b. tissue capillaries
c. normally does not occur anywhere

> LO 11: Describe the natural defenses of the lungs
> LO 12: Describe the cause, effects, and treatment of lung disorders such as:
>   a. Pneumonia   c. Asthma
>   b. Tuberculosis d. Abestosis and silicosis
> LO 13: Summarize the health effects of smoking, with emphasis on the respiratory system.
>
> LO 14: Give the sources, nature, and principal health effects of photochemical and SOP smog.

The cilia-mucus escalator is rendered more effective by inducing turbulance into the air stream and slowing it down, for example, by bronchial constriction. In addition to the cilia-mucus escalator mechanism, the air ducts are served by secreted immuno-globulins, macrophages, and granulocytes. Leukocytes and macrophages are practically the only defense possessed by the alveoli against invasion by particles or bacteria.

Pleurisy is a painful inflammation of the pleural membranes separating the lungs from the chest wall. COPR involves the destruction of alveoli and the obstruction of the smaller air passages with mucus. Asthma obstructs the small air passages with mucus and by muscular spasm. Asbestosis, silicosis and related diseases damage the alveoli by the mechanical and chemical action of inhaled particles. Lung tissue is destroyed by tuberculosis, a bacterial infection, but not usually by pneumonia which simply fills the alveoli with fluid and has a variety of causes.

Air pollution and smoking have similar effects, both causing COPD and, especially in the case of smoking, lung cancer. Smoking is also known to contribute to such diseases as cardiovascular disorders and gastric ulcers. Both smoking and air pollution aggravate other respiratory disease.

SOP smog results from industrial and power-generation combustion of coal and oil; photochemical smog from the action of sunlight upon hydrocarbon fumes and NO_x, both produced mainly by transportation sources especially automobiles.

(1) How is the effectiveness of the cilia-mucus escalator increased by:
   a. irregularities in the respiratory structure that produce air turbulence_____

   b. reduction of the size of the lumens of the respiratory ducts_____

(2) How does congestive heart failure cause pulmonary edema?_____

(3) Briefly describe:
   a. COPD_____
   b. Asthma_____
   c. Pneumonia_____
   d. Tuberculosis_____

e. Silicosis_____
   f. Asbestosis_____
   g. Pleurisy_____
   h. Sinusitis_____
(4) Give four health reasons why one should not smoke:
   1._____
   2._____
   3._____
   4._____

TEST YOURSELF

28. Which of the following is not a normal defense mechanism of the respiratory system   a. bronchial constriction   b. mucus secretion   c. ciliary action   d. alveolar edema   e. local production of antibodies

29. A major source of SOP smog is   a. power generation   b. automobiles   c. trash burning   d. street dust   e. evaporation of paint and other solvents

30. Caused almost exclusively by smoking
31. Alveoli fill with fluid
32. Particles kill macrophages
33. Allergic constriction of bronchi and bronchioles
34. Destruction of alveoli, causing them to coalesce in combination with chronic bronchitis

   a. COPD
   b. lung cancer
   c. pneumonia
   d. asthma
   e. silicosis

> *LO 15: Summarize the mechanics of quiet and forced breathing, giving the contribution of each involved structure or set of structures.*

In quiet breathing, inspiration is active, resulting from the enlargement of the chest cavity by the action of the external intercostal muscles and the diaphragm. Expiration is passive. In forced respiration accessory muscles are brought into play, especially in expiration.

(1) Why does air enter the lungs during inspiration? Explain:_____
(2) Why does air leave the lungs during expiration? Explain:_____

(3) How does forced breathing (for example, during public speaking or exercise) differ from quiet breathing?_____

(4) List some of the accessory respiratory muscles that may be brought into action during dyspnea (see Table 7-8)._____

(5) What is the function of the pleural fluid?_____

(6) What is pneumothorax?_____

(7) Pulmonary disease generally alters the volume indices of respiratory function. After studying Table 15-2 list and define the volume indices as there described.
   a._____
   b._____
   c._____
   d._____
   e._____
   f._____

TEST YOURSELF

35. The volume of the parts of the trachea, bronchi and other structures which do not participate in gas exchange is known as   a. tidal volume   b. residual volume   c. dead space   d. vital capacity   e. expiratory reserve volume

36. When the diaphragm contracts   a. the size of the chest cavity increases   b. the lungs expand to fill the extra space in the chest cavity   c. air from outside the body rushes into the lungs   d. answers (a),(b) and (c) are correct   e. only two of the preceding answers are correct

37. Which of these muscles are _expiratory_ in action   a. diaphragm   b. internal intercostals (in part)   c. external intercostals (in part)   d. tensor fascia lata   e. trachealis

38. The pleural fluid serves to   a. lubricate the apposed membranes of the lungs and chest cavity   b. prevent pneumothorax   c. entangle particulate matter and bacteria   d. moisten the exposed surfaces of the alevoli   e. attack bacteria with antibodies

> LO 16: a. Summarize the neural control of respiration in relation to the pH, CO_2 and O_2 content of the blood, summarizing the chemical and nervous mechanisms involved.
>   b. Describe the role of the respiratory system in the control of blood pH.

Reduction of the blood pH is sensed by the aortic and carotid bodies, and eventually, by the medulla itself. Usually such acidosis is associated with increased carbon dioxide content of the blood, and the medullary respiratory center responds by increasing respiration. The medullary respiratory center actually responds to a reduction in CSF pH resulting from blood acidosis, but if the acidosis is long continued, CSF pH is increased and respiration is controlled by blood oxygen content.

(1) The medullary respiratory center stimulates _____, and is then inhibited, so that _____ takes place passively.
(2) Several sets of receptors seem to influence the action of the respiratory. List them:
   a._____
   b._____
   c._____
   d._____
(3) How does the respiratory pacemaker adapt to chronic acidosis?_____
(4) What then comes to govern respiration?_____
(5) Why must oxygen be administered only with great caution to victims of chronic respiratory disease?
   _____
   _____

(6) What are the physiologic results of hyperventilation?_____
   _____

(7) To sum up, how does the respiratory system prevent excessive reduction of blood pH? _____

TEST YOURSELF

39. The respiratory pacemaker is stimulated by oxygen deficit   a. as its normal mode of control   b. in hyperventilation   c. in chronic acidosis   d. when the CSF pH becomes excessively low   e. in none of the above

ANSWERS TO ALL TEST YOURSELF QUESTIONS

1.a   2.c   3.b   4.c   5.d   6.e   7.b   8.d   9.c
10.d* 11.a   12.c   13.c   14.c   15.c,d 16.a   17.b
18.c   19.a   20.c   21.c   22.a   23.c   24.b   25.a
26.b   27.c   28.d   29.a   30.b   31.c   32.e   33.d
34.a   35.c   36.d   37.b   38.a   39.c

POST TEST   (Nos. 1-5 refer to LO 1)

1. bronchi   a. contains vocal cords
2. nasal cavity   b. lymphatic structures that protect
3. alveoli   respiratory system from infection
4. larynx   c. terminal air sacs
5. tonsils   d. humidifies air
                    e. small air tubes with mucus glands
                       and cartilage

6. (LO 3-5) The sinus cavities   a. produce mucus   b. are located in skull bones   c. communicate with the nasal cavities   d. are of largely unknown function   e. all of the preceding

* The larynx and glottis close whenever high intra-abdominal pressure is produced by the contraction of the abdominal muscles. Thus, strange as it may seem, the larynx does function in defecation.

7. (LO 3-5) The nasal cavity a. is divided by a medial septum b. dries the inhaled air c. captures particulate matter in special cavities d. is connected to the middle ears by the eustachian tubes e. is fringed with coarse hairs near the internal nares

8. (LO 6) Which of these is the innermost a. nasopharynx b. oropharynx c. laryngopharynx d. internal nares e. soft palate

9. (LO 7) In administering mouth-to-mouth resuscitation you should a. place the victim on his back and clear his mouth and throat b. tilt the victim's head back and pinch his nostrils shut c. blow vigorously into his mouth about twelve times each minute d. check to be sure that the air rushes back out of his lungs e. do all of the preceding

NOTE: CPR is a matter of literally life and death importance. You owe it to yourself and everyone else to be familiar with every detail of the procedure.

10. (LO 8) The bronchi possess a. the trachealis muscle b. pleural fluid glands c. a supporting webwork of fibrous connective tissue d. C-shaped rings of cartilage e. walls only one cell in thickness

11. (LO 9) Which of these can the lung dispense with if necessary (as in surgery)? a. the alevolar capillaries b. the bronchial circulation c. the respiratory bronchioles d. the bronchi e. the bronchial mucus

12. (LO 10) The Bohr effect directly causes a. CO_2 to diffuse out of the pulmonary capillaries b. Cl^- to be exchanged for HCO_2^- in the red blood cells c. oxygen to combine with hemoglobin d. oxygen to be liberated from homoglobin e. the production of carbonic anhydrase

13. (LO 10) When oxygen diffuses into the blood, it
    a. forms a strong chemical bond with hemoglobin
    b. forms a weak chemical bone with hemoglobin
    c. it dissolved in the plasma and transported principally in this fashion   d. combines with carbon dioxide to form a special type of transport molecule   e. does two of the above

14. (LO 10) Oxygen diffuses from the air sacs into the blood because   a. carbon dioxide is more concentrated in the air sacs   b. carbon dioxide is less concentrated in the air sacs   c. oxygen is more concentrated in the air sacs   d. oxygen is less concentrated in the air sacs   e. two of the above are correct

15. (LO 11) Particulate matter which invades the lungs may be   a. trapped in mucus   b. tackled by macrophages   c. transported to lymph nodes   d. stored in the nasal conchae and crypts of the tonsils   e. (a),(b) and (c) are correct

(Nos.16-18 refer to LO 12)
16. Alevolar incompetence that develops gradually in company with obstruction of air ducts as a result of respiratory system abuse
17. Temporary response of the respiratory system to irritation
18. Wild multiplication of bronchial lining cells produced almost exclusively by smoking

   a. pneumonia
   b. COPD
   c. pleurisy
   d. bronchogenic lung cancer
   e. bronchial constriction

19. (LO 13) Which of the following is a pollutant occuring in cigarette smoke that interferes with the ability of the blood to transport oxygen?   a. nicotine   b. carbon monoxide   c. asbestos   d. hydrogen cyanide   e. hydrocarbons

20. (LO 14) Which combination would be likeliest to produce SOP smog exclusively?   a. heavy traffic, cloudy weather   b. heavy traffic, sunny weather   c. drought and high winds   d. heavy industry, few automobiles   e. impossible to predict

21. (LO 15) When the diaphragm and rib muscles relax a. the size of the chest cavity decreases b. expiration occurs c. the lungs expand to fill the extra space d. answers (a),(b) and (c) are correct e. only answers (a) and (b) are correct

22. (LO 16) During exercise a. greater amounts of carbon dioxide are produced b. the respiratory center in the brain is stimulated by the carotid and aortic bodies c. nerves from the respiratory center stimulate contraction of chest muscles d. the pH of the CSF may decline e. all of the preceding

ANSWERS TO POST TEST QUESTIONS

1. e   2. d   3. c   4. a   5. b   6. e   7. a   8. c   9. e
10. d   11. b   12. d   13. b   14. c   15. e   16. b   17. e
18. d   19. b   20. d   21. e   22. e

16 THE URINARY SYSTEM

1. The kidneys are located _____ the peritoneum in the _____ part of the _____ cavity.
2. The outer "rind" of each kidney is known as its _____. The substance of the kidney is divided internally into _____, each of which terminates in an apex called _____, the very tip of which, the _____ contains pores through which urine drains into the _____.
3. The functional units of the kidneys are the _____.
4. Bowman's capsule of each nephron functions to _____ the blood.
5. Compare the cortical and juxtamedullary nephrons by completing this table:

Structure	cortical	juxtamedullary
Loop of Henle	moderate	very long
Capillary network	extensive	a)
Location	b)	c)
Vasae rectae	absent	d)

6. This is a diagram of a cortical/juxtamedullary nephron. Label the blanks.

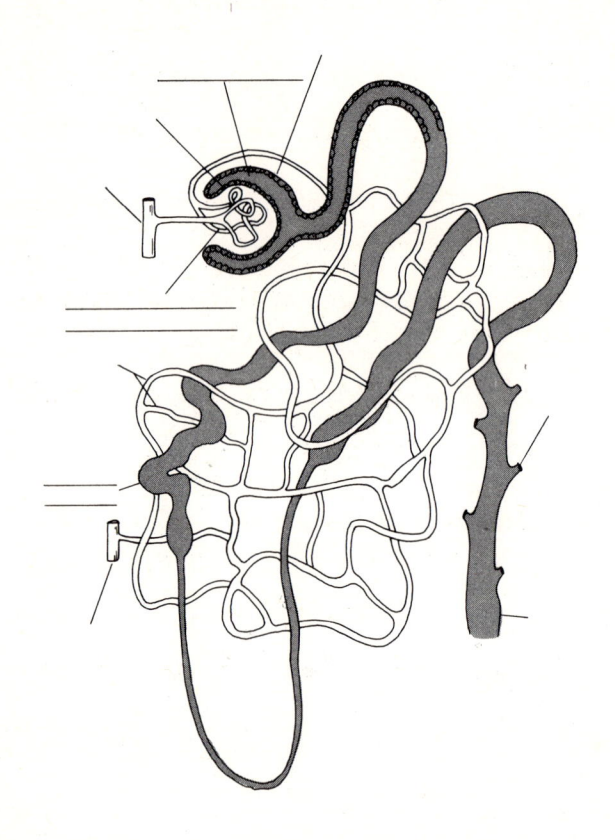

7. The glomerulus is an expanded portion of the _____ arteriole whose walls are composed of endothelium provided with _____. The cells of the wall of Bowman's capsule are highly specialized and are known as _____. Their foot processes appear to regulate the passage of fluid into the nephron lumen.
8. Filtration is relatively unselective, so that most of the difference between blood plasma and urine is produced by _____.
9. In the case of a cortical nephron, the _____ _____ of the _____ _____ cells transports sodium into the _____. Water follows by _____.

10. Glucose and protein are reabsorbed by the_____
    _____. Chloride is reabsorbed by the
    _____.
11. What is meant by renal threshold?_____
    _____
12. The sodium pump is able to move sodium from an area
    of_____concentration to one of_____
    concentration, thus opposing ordinary principles of
    diffusion.
13. The filtrate leaving the nephron is, generally
    speaking,_____-tonic to blood though much
    reduced in volume. However, the loops of Henle of
    the_____nephrons establish a zone of
    _____-tonicity that draws off water from the
    urine in the_____.
14. ADH_____the permeability of the collecting
    duct, thus_____the volume of urine and also
    the tonicity of the blood by_____the
    proportion of water that is reabsorbed.
15. Sodium reabsorption is governed principally by the
    hormone_____which is secreted by the_____
    _____.
16. Blood volume appears to be regulated by the actions
    of two hormones_____and_____.
17. In acidosis, the kidney secretes_____into
    the tubules. This may combine with_____, or
    if sodium must be conserved due to extremely low
    blood pH, with_____.
18. Casts appearing in the urine always indicate some
    kind of_____, but bacteria are significant
    only if the specimen is absolutely _____.
19. The principle drawback of kidney transplantation
    is the possibility of graft_____. Dialy-
    sis, on the other hand, is inconvenient and very
    _____.
20. The ureters are posterior, retroperitoneal tubes
    stretching from the renal_____to the
    _____. They pump urine by a_____
    action.
21. The urinary bladder is lined with_____
    epithelium. Its muscle layer is called the
    _____muscle.
22. The urethra connects the_____with the
    outside world. The urethra possesses___sphincter
    valves, the external one composed of_____

muscle. The _____ reflex opens these sphincters and empties the bladder.

ANSWERS TO PRETEST QUESTIONS

1. behind--posterior--abdominal (LO 1) 2. cortex--lobes--pyramid--papilla--renal pelvis (LO 1) 3. nephrons (LO 2) 4. filter (LO 2) 5. a) minor b) outer portion of organ c) inner portion of organ, near pyramids d) present (LO 2) 6. compare with appropriate figures in the textbook (LO 2) 7. afferent--pores--podocytes (LO 3) 8. reabsorption (LO 4) 9. sodium pump--proximal convoluted tubule--interstitium--osmosis (LO 4,5) 10. proximal convoluted tubule--distal convoluted tubule (LO 4) 11. blood concentration of a substance over which not all of it can be absorbed (LO 4) 12. lesser-greater (LO 5) 13. iso--juxtamedullary--hyper--collecting ducts (LO 6) 14. increases--reducing--increasing (LO 7) 15. aldosterone--adrenal cortex (LO 7) 16. ADH--aldosterone (LO 7) 17. H^+--$NaHPO_4$--NH_4^+ (LO 8) 18. disease--fresh (LO 9) 19. rejection--expensive (LO 10) 20. pelves--bladder--peristaltic (LO 11) 21. transitional--detrusor--contracts (LO 12) 22. bladder--two--voluntary--micturation (LO 13, 14)

LO 1: Describe the location and gross structure of the kidney, and give its general functions.

The kidney is a bean-shaped organ which internally consists of a cortex, a medulla, and a pelvis. The cortex and medulla are divided into lobes, each of which drains separately into the pelvis via pyramids and papillae. Ureters drain the pelvis, and each ureter is an extension of the pelvis. Urine is formed in the nonpelvic portions of the kidney. The kidneys function both in the excretion of waste products and in otherwise maintaining the homeostasis of the body fluids.

(1) The kidneys do, of course function in the _____ of waste products, but their other homeostatic functions are probably even more important.

(2) Label the following diagram:

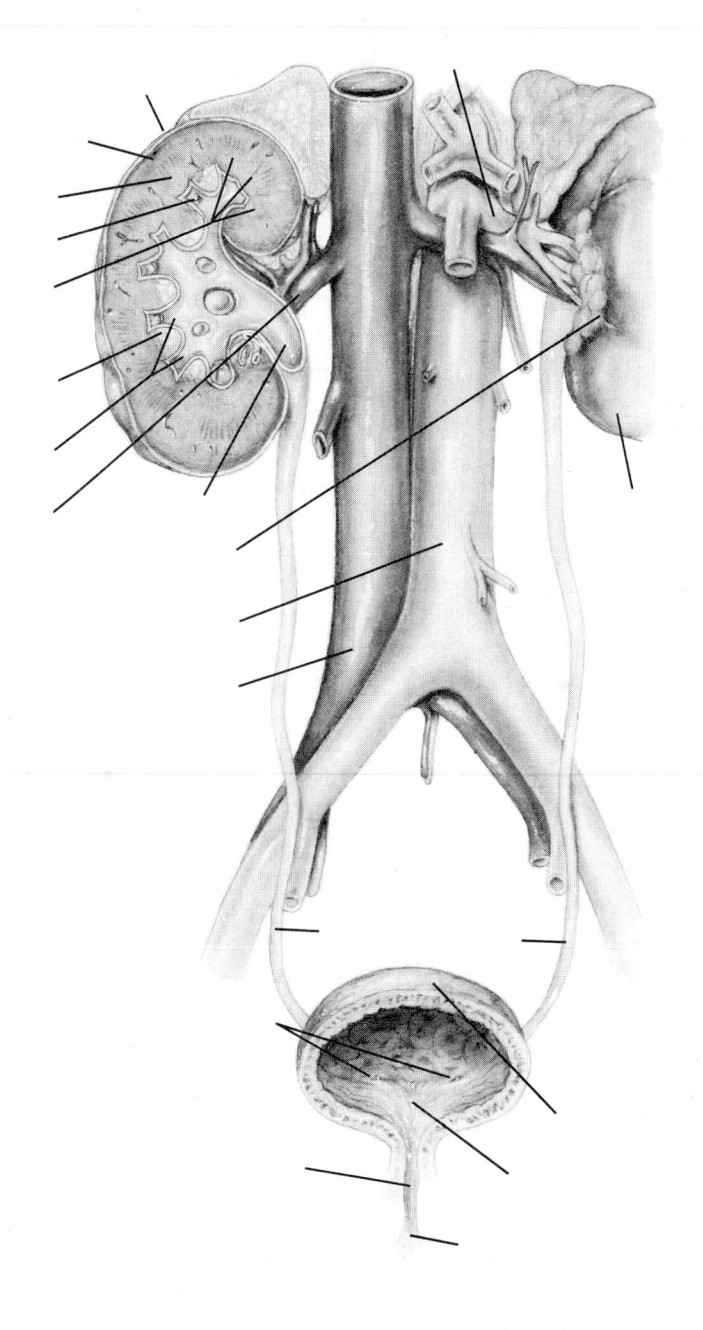

(3) Glance back at table 1-1, then mentally review the entire chapter. Then list the ways (other than waste product excretion) in which the kidneys help maintain body homeostasis.
1._____
2._____
3._____

TEST YOURSELF

1. The kidney is located a. in the abdominal cavity b. behind the posterior peritoneum c. beneath the last two pairs of ribs d. in the pelvic basin e. (a),(b) and (c) are correct

2. The kidney functions in a. disposal of nitrogenous wastes (such as urea and uric acid) b. maintainance of blood pH c. maintainance of blood volume d. maintainance of blood osmotic pressure e. all of the preceding ways.

> LO 2: Describe the main types of nephrons in detail, distinguishing between the cortical and juxtamedullary nephrons with respect to proportions, location, blood supply and function.

The basic excretory unit of the kidney is the nephron, of which there are about a million in each kidney. In general, nephrons consist of the capillary tuft or glomerulus, Bowman's capsule, the proximal convoluted tubule, the loop of Henle, and the posterior convoluted tubule. There are two types of nephrons; the cortical and juxtamedullary. Though both types of nephrons seem to function in excretion, only the juxtamedullary nephrons are involved in the production of hypertonic urine.

(1) Urine is produced by the kidney_____, whose basic unit is the _____.
(2) What are the most important structural differences between the cortical and juxtamedullary nephrons?
_____
_____
_____

(3) In the space below, diagram a typical nephron, labelling the parts:

TEST YOURSELF

3. Which of the following structures contains blood?
   a. glomerulus   b. Bowman's capsule   c. convoluted tubules   d. vasae rectae   e. (a) and (d)

4. Juxtamedullary nephrons   a. are located in the kidney cortex   b. have an exceptionally large capillary network   c. have an exceptionally long loop of Henle   d. are responsible for the production of isotonic urine

> LO 3: Describe the role of the glomerulus and of Bowman's capsule in filtration.

Modified blood plasma filters through the glomerulus into Bowman's capsule. This filtrate forms the basis of urine. Filtration is, for the most part, not *selective. However, most of the plasma protein remains behind in the blood.*

(1) The renal corpuscle consists of the _____ and _____.

(2) Blood in the _____ arteriole is under fairly high hydrostatic pressure.

(3) Look ahead to the section REGULATION OF BLOOD PRESSURE. How does the efferent arteriole regulate blood pressure in the glomerulus? _____
_____

(4) The endothelium of the glomerulus contains many _____. These are partly closed off by the _____ processes of the _____ cells of Bowman's capsule. It is believed that these structures _____ the passage of materials into the filtrate.

(5) The filtrate is essentially blood plasma, but minus its _____.

(6) What is the significance of renal clearance? _____
_____

(7) What is the general procedure for administering a renal clearance test? _____
_____
_____

TEST YOURSELF

8. Podocytes: a. occur in the glomerulus b. appear to regulate filtration c. engage in the active transport of Na^+ d. have brush borders e. line the collecting duct

6. The blood in the efferent arteriole a. is similar in composition to ordinary arterial blood except that most waste products have been removed b. is under somewhat higher hydrostatic pressure than that in the afferent arteriole c. is hypertonic to afferent arteriolar blood d. is low in plasma protein e. is high in glucose and most salts

> LO 4: Recount the role of the nephric tubule in the concentration, reabsorption and secretion of the various substances important to its function, especially water, sodium, urea and glucose.
>
> LO 5: Summarize the operation of the sodium pump, and describe the role it plays in the reabsorption not only of sodium but also of water in the kidney.

The proximal convoluted tubule selectively reabsorbs most sodium and water from the urine. In the case of the juxtamedullary nephrons, the loops collectively establish a zone of interstitial hypertonicity that, by producing water absorption from the collecting ducts, can produce a hypertonic urine.

Water, however, is only indirectly reabsorbed and is not itself actively transported. The sodium pump functions in the proximal convoluted tubule and (in the case of juxtamedullary nephrons) in the loop of Henle by transporting sodium out of the tubule. Water then follows passively by osmosis.

Some substances, such as glucose, have a critical blood concentration below which they are completely reabsorbed and so they do not appear in the urine until this _threshold_ value is exceeded.

(1) What is obligatory reabsorption?_____

(2) Where in the cortical nephron does it appear to take place?_____

(3) The reabsorption of water in the nephron itself is a direct/indirect process. The brush-border epithelial cells are specialized for the transport of_____. Negative ions accompany this into the _____ fluid of the kidney. _____ follows by osmosis.

(4) All this produces a(n) isotonic/hypotonic fluid at the terminus of the distal convoluted tubule.

(5) In the case of juxtamedullary nephrons, though their product is isotonic/hypotonic/hypertonic to blood, the loop of Henle produces _____ conditions in the part of the interstitium through which all collecting ducts must pass.

(6) Thus, the collecting ducts, which are permeable to _____ but not to most solutes, lose _____ to the interstitium by osmosis and are capable of producing isotonic/hypotonic/hypertonic urine.

(7) What is the role of the vasa rectae? _____
_____
(optional) why does it seem logically necessary that they possess active sodium pumps? _____

(8) Summarize the reabsorption of such substances as glucose and amino acids here: _____
_____

(9) What role does secretion play in the kidney? _____
_____

TEST YOURSELF

FLUID	TONICITY (COMPARED TO ORDINARY BLOOD)
7. nephron drainage	
8. fluid within the base (lowest point) of the loop of Henle of a juxtamedullary nephron	a. isotonic b. hypotonic
9. fluid draining from papillary openings of the collecting ducts during thirst	c hypertonic
10. juxtamedullary interstitial fluid	

11. Water leaves the nephron   a. by active transport   b. by filtration into the capillary network   c. by osmosis   d. by the direct action of active transport mechanisms   e. (a) and (d)

12. Where would the greatest volume of fluid be found?
    a. Bowman's capsule   b. proximal convoluted tubule
    c. loop of Henle   d. distal convoluted tubule
    e. collecting duct outlet

13. Where would the <u>least</u> volume be found? (Use the above choices):

> *LO 6: Summarize the role of the collecting ducts in the reabsorption of water.*
>
> *LO 7: Describe the action of ADH and aldosterone on the nephron and collecting duct, and summarize their role in body fluid homeostasis.*

As mentioned earlier, the hypertonicity of the juxtamedullary interstitium though which all collecting ducts must pass on their way to the renal pelvis draws water out of them by osmosis. But except for some urea which is reabsorbed here to a small extent, most solutes are left in the ducts, thus producing a hypertonic urine. But this production of hypertonic urine will only occur if the collecting ducts are fully permeable to water, so the regulation of urine tonicity depends upon the control of the permeability of the collecting ducts to water, a function of ADH. Aldosterone regulates the action of the sodium pump in at least the proximal convoluted tubule, and therefore the reabsorption of sodium.

Working together, ADH and aldosterone govern blood volume.

<u>NOTE</u>: The advantages of this very elaborate-appearing setup may still not be obvious to the student, and indeed they appear to be unclear to many persons more knowledgeable. To our minds, though, there are two advantages which taken together may account for many of the eccentricities of kidney function:
  1) only by using some kind of countercurrent exchange (as in the juxtamedullary loops of Henle) can the body produce hypertonicity in an excreted product without exceeding the capacity of the sodium pump to transport sodium across a tonicity difference at any one point, and
  2) since there is no known mechanism for the active transport of water, the excretion of water can only be regulated by regulating the active sodium pump or the passive reabsorption of water. Evi-

dently it is more practical to regulate the reabsorption of water, for if it were to be done by regulating the sodium pump one would have no way of producting hypertonic urine or for that matter hypotonic urine at need. Only urine volume could be regulated. As it is, urine volume and tonicity are both regulated by restricting the reabsorbtion of water from the collecting ducts. In this way either sodium or water can be conserved, as may be appropriate.

(1) ADH governs the permeability of the _____ _____ wall to _____.
(2) ADH is secreted by the _____ _____ gland in response to blood hypo/hyper-tonicity sensed by special sense organs in the brain.
(3) One might think that diabetes insipidus would be immediately fatal because of loss of the total water of the glomerular filtrate could not be replaced by any amount of drinking. Can you propose an explanation involving the proximal convoluted tubule of why this does not occur, and that such a sufferer is able to keep alive, however inconveniently, by great water intake?

(4) As ADH increases, duct wall permeability_____, Therefore water reabsorption_____, so urine volume_____.

(5) Imagine that you have eaten a package of salted peanuts. This makes you thirsty. You drink a glass of water. In the space below describe the complete sequence of events whereby you get rid of the salt and water and avoid both high blood pressure and excessive blood volume.

(6) Disbetes insipidus is produced by_____ deficiency. An untreated sufferer must_____

(7) The operation of the kidney sodium pump appears to be governed by the hormone_____, which is produced by the_____ _____.

TEST YOURSELF

14. ADH governs   a. the sodium pump of the proximal convoluted tubule   b. the pressure of the blood in the efferent arteriole   c. the sodium pump of the vasae rectae   d. the water permeability of the lower portions of the loop of Henle

15. Excess intake of water normally leads to   a. hypertonicity of the blood   b. reduced ADH secretion   c. changes in aldosterone secretion   d. increased permeability of the collecting ducts to water   e. increased blood volume

> LO 8:   Summarize the homeostatic roles of the kidney in maintaining blood pH and blood pressure and describe how the nephron accomplishes this.

Both the kidney and the lung cooperate to maintain blood pH homeostasis. The kidney is able to excrete H^+ and to neutralize it partly with the aid of NH_3 which it manufactures from amino acids. In addition, the kidney is able to conserve bicarbonate and sodium by a series of exchange reactions.

Falling blood pressure stimulates the release of renin by the juxtaglomerular cells of the afferent arteriole which in turn increases the systemic blood pressure of the entire body. The filtration of blood is dependent upon the pressure of blood in the glomerulus. This is maintained constant on the glomerular level by continual adjustment of the diameters of the efferent glomerular arterioles. As you should also recall, blood pressure is regulated by the kidney through its ability to govern blood volume.

(1) Under conditions of excessively low blood pH , the kidney tubules excrete_____. This, in turn, is buffered in urine by_____.

(2) If blood pH is extremely low indeed,_____ is produced by amino acid catabolism, and this, instead of sodium is employed to_____ urinary H$^+$

(3) What is the role of HCO_3^- ?_____

(4) How is high blood pH countered by the kidney?_____

(5) Summarize how, through hypo or hyperventilation, the <u>respiratory</u> system helps to regulate blood pH.

(6) The distal convoluted tubule is adjacent to the _____ arteriole, which possesses at that point a cuff of_____ cells which is sensitive to both blood_____ and to tubular concentration of _____. These cells respond to low blood pressure by secreting_____, which raises it systemically. Locally, glomerular blood pressure is also regulated by the constriction of the_____ _____.

TEST YOURSELF

16. When blood pH is unusually <u>low</u>, the kidney
    a. secretes H$^+$ b. reabsorbs less Na$^+$ than usual
    c. reabsorbs NH$_3^+$ d. makes less carbonic anhydrase   e. (b) and (c)

309

17. Which of the following is NOT involved in the regulation of blood pressure by the kidney?
a. juxtaglomerular cells   b. renin   c. low tubular sodium content   d. efferent arteriole   e. all of the preceding are in fact involved

> LO 9: Be able to give specific examples of the use of the urine in clinical diagnosis.

Study table 16-1, then describe typical findings in the following disease states:(1 and 2) or the disease significance, if any, of the findings (3 and 4)

1. Diabetes mellitis_____
2. Diabetes insipidus_____
3. Squamous epithelial cells_____
4. Red blood cell casts_____

TEST YOURSELF

URINALYSIS FINDINGS   POSSIBLE DIAGNOSIS

18. Uric acid crystals   a. diabetes insipidus
19. Low specific gravity   b. bacterial infection
20. Glucose   c. terminal kidney disease
21. High pH (fresh specimen)   d. gout
22. waxy casts   e. diabetes mellitis

> LO 10: Summarize the treatment of renal failure, with special reference to artificial kidney dialysis and to kidney transplantation.

Renal failure, if complete and permanent, must be treated by some kind of replacement of the kidney, whether by dialysis or transplantation of a live kidney. Live kidneys, however, are subject to immunologic rejection. Dialysis, on the other hand, is inconvenient, often painful, and ruinously expensive.

(1) What is artificial kidney dialysis?_____
_____

(2) What difficulties surround the use of transplanted kidneys? (a) donor choice_____
    (b) graft rejection_____

310

TEST YOURSELF

23. The uremic syndrome a. involves disturbances of blood pH and solute balance b. usually results from the removal of a single kidney c. cannot be treated by dialysis d. greatly increases the chance that a grafted kidney will be rejected

>LO 11: Describe the structure, location and function of the ureters.

Two ureters connect the bladder with the kidney. Periodic peristalsis of the ureters delivers urine into the bladder.

(1) Ureters are tubular extensions of the _____ _____.
(2) The _____ lining of the ureter protects it from the action of the urine it conveys.
(3) _____ delivers periodic jets of urine to the bladder. When this is full, the _____ can themselves function as accessory bladders.

TEST YOURSELF

24. The ureters are a. located under the posterior abdominal peritoneum b. tubular extensions of the renal pelvis c. the main outlet of the bladder d. lined with transitional epithelium e. <u>three</u> of the preceding are true

>LO 12: Describe the structure and summarize the function of the urinary bladder and urethra.
>LO 13: Compare the course and structure of the urethra in males with that in females.
>LO 14: Outline the micturition reflex with special reference to the operation of the urethral sphincter and its inhibition.

The urinary bladder consists of three layers: the inner transitional epithelium, the middle cross-grained detrusor muscle, and the outer serosa. The epithelial lining of the bladder is specially adapted to stretching. The urethra is short in women, relatively long in men. The male urethra, which must serve as a passage for both seminal fluid and urine, is divided into penile, membranous, and prostatic portions. It is provided with two sphincters at its attachment to the bladder. These must open, and the bladder must contract, to produce micturition. The events are initiated by spinal reflex when the bladder becomes turgid. The micturition reflex can be consciously both inhibited and facilitated.

1. Label the following diagram:

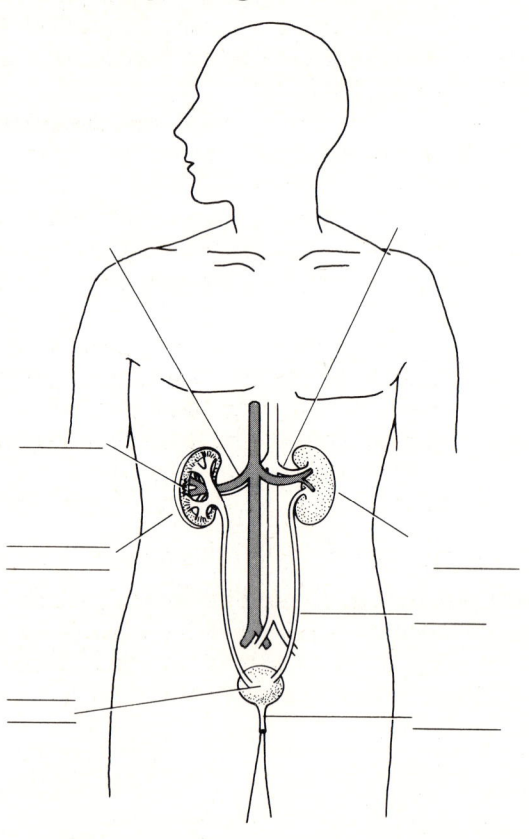

(2) The bladder wall consists of four layers (begin with the innermost): 1._____
    2._____ 3,_____
    4._____
(3) Transitional epithelium is remarkable for its ability to _____
(4) During the micturition reflex, the_____ _____contracts and the urethral_____ open.
(5) The urethra is longest in males/females and in them is subdivided into three portions: 1._____
    2._____ 3._____

(6) Flaps of tissue located at the openings of the
    ureters into the bladder normally prevent_____
    of urine even when the bladder is fully distended.

TEST YOURSELF

25. lined with transitional epi-   a. urinary bladder
    thelium   b. urethra
26. operates by perist lsistalsis   c. ureters
27. contains detrusor muscle   d. (a) and (c)
28. longer in males than in females   e. (b) and (c)
29. openings of the trigone
30. becomes infected in cystitis

ANSWERS TO ALL TEST YOURSELF QUESTIONS

1.e   2.e   3.e   4.c   5.b   6.c   7.a   8.c   9.c
10.c   11.c   12.a   13.e   14.e   15.b   16.a   17.e
(and, one might add, ADH and aldosterone too!)
18.d   19.a   20.e   21.b   22.c   23.a   24.e   25.d
26.c   27.a   28.b   29.e   30.a

POST TEST

1. (LO 1) The functions of the kidneys are   a. to help
   maintain the internal chemical balance of the body
   b. storage of urine   c. producing urine   d. all
   of the preceding are correct   e. two of the preceding are correct

2. (LO 2) Hypertonic urine is produced as a result of
   the action of the a. cortical nephrons   b. proximal
   convoluted tubules   c. loops of Henle (of the juxtamedullary nephrons)   d. bladder   e. ureters

3. (LO 3) A portion of the blood is filtered as it
   passes through the wall of the   a. excretory tubule
   b. renal vein   c. posterior convoluted tubule
   d. renal artery   e. ureter

4. (LO 3) Glomerular filtration may be said to   a. produce hypertonic urine   b. produce hypotonic fluid
   c. produce hypertonic blood   d. be highly selective
   e. be the function of the brush-border cells

5. (LO 4) In the condition diabetes mellitis a. an insufficient amount of insulin is secreted b. carbohydrate metabolism is impaired c. the concentration of glucose filtered into the nephrons exceeds the renal threshold d. statements (a),(b) and (c) are correct e. two of the preceding statements are correct

6. (LO 4) When a substance exceeds its renal threshold, a. excess insulin is secreted b. death can occur swiftly c. the portion not reabsorbed is excreted in the urine d. the excess is returned to the blood by the process of secretion e. two of the preceding answers are correct

7. (LO 4) During urine formation a. useful materials are returned to the blood by the process of reabsorption b. wastes and excess materials are secreted into the renal veins c. an adjusted filtrate consisting primarily of sugar and urea is formed d. all of the above are correct e. only two of the preceding answers are correct

8. (LO 4) Certain drugs such as penicillin are excreted primarily a. by the sweat glands b. through the lungs c. by the process of tubular secretion d. by filtration e. as a result of renal threshold

9. (LO 4) Renal threshold is a. the amount of glucose which appears in the urine of a diabetic b. an insufficiency in the amount of insulin secreted c. an insufficiency in the amount of ADH secreted d. the maximum concentration of a specific substance in the blood at which complete reabsorption can take place e. the amount of a substance which can be filtered by Bowman's capsule

10. (LO 4) Which of the following could be called a highly selective process a. filtration b. reabsorption c. passage of plasma into Bowman's capsule d. all are selective e. two of the preceding are selective

11. (LO 4) Reabsorption of useful materials takes place a. through Bowman's capsules b. through the capillary masses in Bowman's capsules c. through the walls of the excretory tubules and collecting ducts d. into the posterior pituitary gland e. as urine passes into the urinary bladder

12. (LO 4) Reabsorption a. is the function of the excretory tubules and collecting ducts b. is a selective process c. takes place through the Bowman's capsule d. takes place in the liver e. answers (a) and (b) are correct

13. (LO 5) _____ is actively transported by the brush-border cells of the proximal convoluted tubule a. Na^+ b. Cl^- c. urea d. H_2O e. (a),(b) and (d)

14. (LO 6) The hypertonic interstitium produced by the _____ draws _____ from the contents of the _____ by osmosis.
    a. cortical nephrons--water--distal convoluted tubules b. collecting ducts--salts--proximal convoluted tubules c. juxtamedullary nephrons--water--collecting duct d. proximal convoluted tubules--HCO_3^---renal corpuscle e. secretory ducts--protein--efferent arterioles

15. (LO 7) When one drinks a large quantity of fluid a. the blood becomes diluted b. ADH release increases c. the amount of water reabsorbed from the collecting ducts is increased d. all of the preceding are correct e. only two of the preceding are correct

16. (LO 7) In the disease diabetes insipidus a. too much insulin is produced b. insufficient amounts of ADH are produced c. a small volume of urine is produced d. statements (a),(b) and (c) are correct e. only two of the preceding answers are correct

17. (LO 7) ADH is secreted by a. the ureters b. Bowman's capsule c. the posterior pituitary gland d. the excretory tubules e. the liver

18. (LO 7) If blood flow through the glomerulus should be reduced   a. the juxtaglomerular cells secrete renin   b. the adrenal cortex secretes aldosterone   c. the posterior pituitary secretes ADH   d. (b) and (c)

19. (LO 9) An increase in blood pH is likely to lead to   a. complete resorption of Na^+ from the tubules   b. increased loss of Na^+ c. secretion of H^+ d. loss of large amounts of HCO_3^-

URINALYSIS OBSERVATION   POSSIBLE DIAGNOSIS

(LO 9 refers to Nos. 20-22)
20. Casts   a. diabetes insipidus
21. Low specific gravity   b. diabetes mellitis
22. Uric acid crystals   c. definite kidney disease
                                    d. acidosis
                                    e. gout

23. (LO 10) In kidney dialysis   a. blood from the patient's body is circulated through a machine   b. an artificial kidney is grafted to the patient's body   c. graft rejection is a major concern   d. statements (a),(b) and (c) are correct   e. statements (a) and (b) are correct

24. (LO 10) In some respects kidney transplants are more advantageous than dialysis for chronic kidney failure because   a. if successful, the total cost is likely to be less   b. uremia is more likely to occur when a successful transplant has been performed   c. a transplant does a better job of removing wastes from the blood   d. statements (a),(b) and (c) are correct   e. only two of the preceding statements are correct

25. (LO 11) Ducts that conduct urine from the kidneys to the urinary bladder are   a. urethras   b. ureters   c. excretory tubules   d. Bowman's capsules   e. nephrons

26. (LO 12) The _____ is a temporary storage sac for urine a. urethra b. serosa c. renal vessel d. kidney e. urinary bladder

27. (LO 13) Which of these structures is markedly different in the two sexes? a. kidney b. ureter c. bladder d. urethra

28. (LO 14) Micturition requires a. a spinal reflex b. contraction of the detrusor muscle c. relaxation of the urethral sphincters d. ureteral peristalsis e. (a),(b) and (c) are correct

ANSWERS TO POST TEST QUESTIONS

1.a   2.c   3.c   4.c   5.d   6.c   7.a   8.c   9.d
10.b 11.c 12.e 13.a 14.c 15.a 16.b 17.c
18.a 19.b 20.c 21.a 22.e 23.a 24.e 25.b
26.e 27.d 28.e

17 ENDOCRINE CONTROL

PRETEST

1. _____ glands have ducts which deliver their secretion, whereas _____ glands simply release their secretion into the surrounding interstitial fluid and capillaries.
2. Endocrine control depends upon chemical messengers called _____.
3. The _____ glands are located above the kidneys; the _____ and _____ are located in the neck.
4. Hormones are generally transported in the _____ chemically bound to _____ _____.
5. The second messenger which has been most extensively studied is _____ _____. It is formed chemically from _____.
6. Hormones are regulated by _____ _____ mechanisms.
7. The _____ is the link between neural and endocrine systems.
8. Hormones released by the posterior pituitary are actually synthesized in the _____.
9. Besides producing tropic hormones, the anterior pituitary produces _____ and _____ _____.
10. Growth hormone acts mainly by promoting _____ synthesis.
11. Hypersecretion of growth hormone in adulthood may result in the condition called _____.
12. Thyroid hormones stimulate the rate of _____.
13. Regulation of thyroid hormone secretion depends mainly upon _____ _____ hormone released by the _____ _____.
14. Hypothyroidism in an adult may result in the condition called _____.

15. _____ hormone stimulates resorption of calcium from bones.
16. Insulin _____ blood glucose level; glucoa-
       (raises, lowers)
    gon _____ blood glucose levels.
        (raises, lowers)
17. In diabetes mellitus ketone bodies accumulate in the blood due to increased _____ metabolism.
18. Two hormones released by the adrenal medulla are _____ and _____.
19. The principal mineralocorticoid is _____.
20. The main function of aldosterone is to increase the rate of _____ reabsorption by the kidney tubules.
21. The principal action of cortisol is to promote _____.
22. Abnormally large amounts of glucocorticoids result in _____ syndrome.
23. In stress the anterior pituitary releases more _____ which increases _____ secretion by the adrenal cortex.
24. Cortisol tends to _____ blood glucose levels.
                      (raise, lower)

ANSWERS: 1. exocrine, endocrine (LO 1) 2. hormones (LO 1) 2. hormones (LO 2) 3. adrenal, thyroid, parathyroids (LO 3) 4. blood, plasma proteins (LO 4) 5. cyclic AMP, ATP (LO 5) 6. negative feedback (LO 6) 7. hypothalamus (LO 7) 8. hypothalamus (LO 7,8) 9. prolactin, growth hormone (LO 8) 10. protein (LO 9) 11. acromegaly (LO 9) 12. metabolism (LO 10) 13. thyroid stimulating, anterior pituitary (LO 11) 14. myxedema (or goiter) (LO 11) 15. parathyroid (LO 12) 16. lowers, raises (LO 13) 17. lipid (fat) (LO 14) 18. epinephrine, norepinephrine (LO 15) 19. aldosterone (LO 16) 20. sodium (LO 16) 21. gluconeogenesis (LO 16) 22. Cushing's (LO 16) 23. ACTH, glucocorticoid (cortisol) (LO 17) 24. raise (LO 17,18).

LO 1: Distinguish between exocrine and endocrine glands.

Endocrine glands are ductless glands that secrete hormones into the interstitial fluid and depend upon the blood to transport them to target tissues.

(1) How do exocrine glands differ from endocrine glands?

(2) Give an example of an exocrine gland _____

TEST YOURSELF

1. an endocrine gland a. lacks a duct b. releases its hormone into the surrounding interstitial fluid c. depends upon the blood for transport of its secretion d. answers (a),(b) and (c) are correct e. none of the preceding answers is correct

> LO 2: Define the term hormone and tell what hormones do:

Hormones are chemical messengers produced by endocrine glands. They regulate the metabolic processes of various cells.

(1) List 4 metabolic activities which are influenced by hormones:
   1. _____
   2. _____
   3. _____
   4. _____
(2) Of what type of organic compounds are hormones composed? _____
   Give 2 specific examples _____

TEST YOURSELF

2. Which of the following are at least partially regulated by hormones a. digestion b. blood glucose level c. metabolic rate d. salt and water balance e. all of the preceding

> LO 3: Identify the principal endocrine glands and locate them anatomically in the body.

Label the endocrine glands in the diagram on the following page:

321

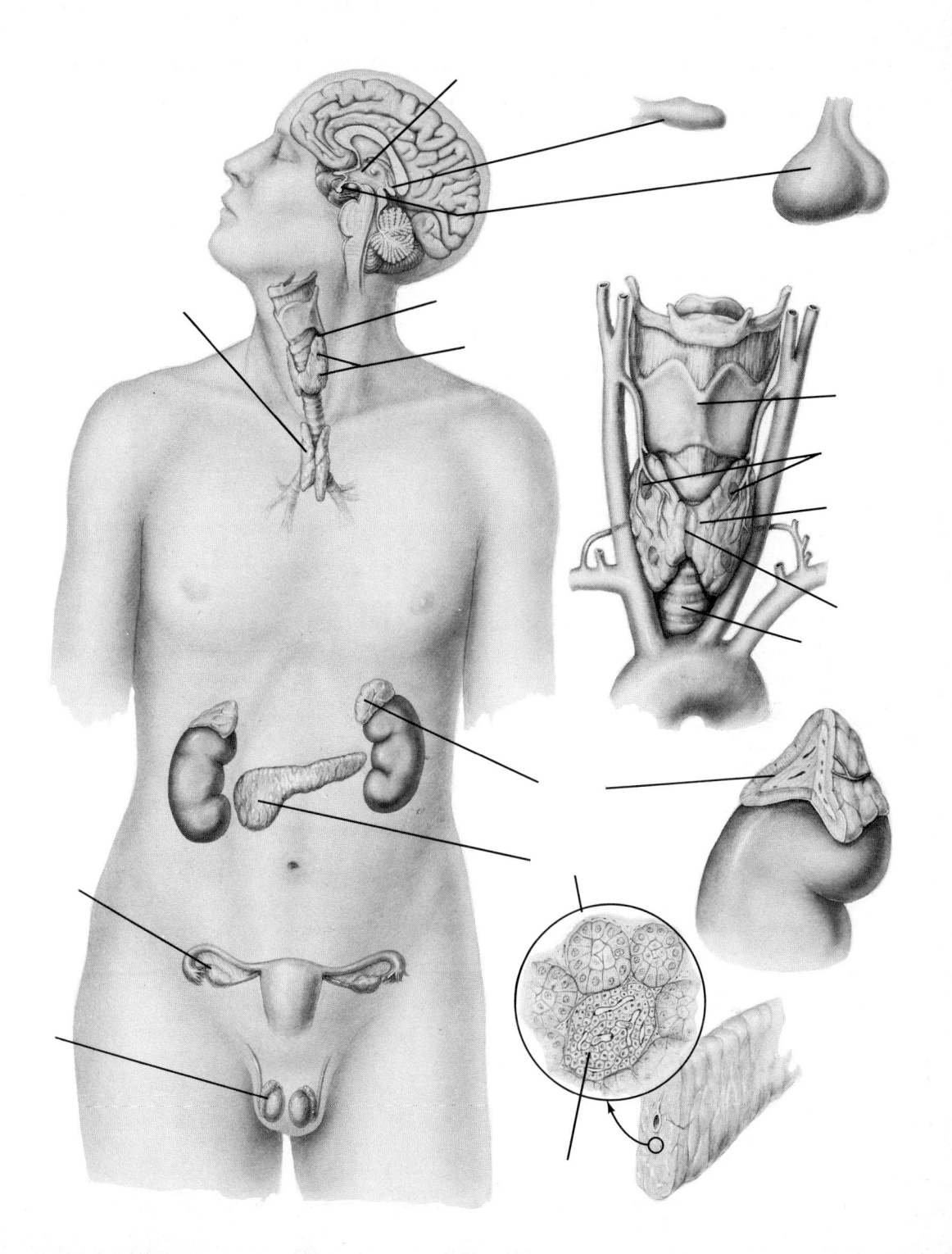

TEST YOURSELF
Match:
3. located in the pancreas   a. anterior pituitary
4. located above the kdineys   b. adrenal glands
5. located in the neck   c. parathyroid glands
6. located in the head   d. islets of Langerhans

> LO 4: Describe the mechanism by which hormone affects its target tissue and describe how hormones are transported.

Hormones are transported in the blood bound to plasma proteins. As needed free hormone is released and finds its way into the tissues. While a hormone may pass by many tissues "unnoticed" as it passes by its target tissue it is bound by specific protein receptors on the cell surfaces, or within the cells.

(1) Use a lock and key analogy to explain the relationship between hormone and target tissue.

(2) Write an equation to illustrate the equilibrium that exists between protein-bound and free hormone in the blood:
_____

TEST YOURSELF

7. Free hormone a. may be found in the blood b. may be removed by the kidneys c. may be found bound to receptors within its target tissue d. answers (a),(b) and (c) are correct e. answers (a) and (c) only are correct

> LO 5: Describe the mechanisms of hormone action including the role of second messengers such as cyclic AMP.

Steroid hormones enter target cells and combine with a receptor in the cytoplasm. The complex then moves into the nucleus and activates specific genes. Nonsteroid hormones may bind to receptors upon the cell membrane of a target cell. A second messenger is activated which actually triggers the chain of events leading to the response.

(1) Tell how cyclic AMP is activated_____

(2) How can cyclic AMP initiate different actions in different types of cells?_____

TEST YOURSELF

8. Cyclic AMP is   a. a second messenger   b. formed from DNA   c. produced when the hormone combines with RNA   d. answers (a),(b) and (c) are correct   e. answers (a) and (c) only are correct

> LO 6: Describe how endocrine glands are regulated by negative - feedback mechanisms: relate to each specific hormone discussed.

Endocrine glands are regulated by negative feedback mechanisms. Too much hormone or too much of the substance regulated by the hormone signals the endocrine gland to slow down.

Draw a simple diagram to illustrate regulation of the parathyroid glands by negative feedback mechanisms.

TEST YOURSELF

9. When the calcium level in the blood rises above normal   a. parathyroid glands produce more hormone   b. parathyroid glands slow their production of hormone   c. parathyroids signal the hypothalamus to slow down   d. the anterior pituitary is inhibited   e. the parathyroids produce less calcium

> LO 7: Tell why the hypothalamus is considered the link between neural and endocrine systems, and describe the mechanisms by which the hypothalamus exerts its control.

The hypothalamus produces several types of releasing and inhibiting hormones which regulate the activity of the *anterior pituitary gland. Hypothalamic cells also produce two hormones, oxytocin and ADH which are released by way of the posterior pituitary gland.*

(1) What is the relationship between hypothalamus and:
   (a) anterior pituitary (adenohypophysis)_____
   (b) posterior pituitary (neurohypophysis)_____

(2) How does the hypothalamus influence other endocrine glands e.g. the thyroid and adrenal medulla.

TEST YOURSELF

10. The hypothalamus a. produces hormones b. produces releasing hormones c. influences glands such as the thyroid and adrenal glands d. answers (a) and (b) only are correct e. answers (a),(b) and (c) are correct

>*LO 8: Identify the hormones released by the anterior pituitary and posterior pituitary, give their origin, and describe their actions.*

Oxytocin and ADH are produced by the hypothalamus but released by the posterior pituitary gland. The anterior pituitary gland produces growth hormone, prolactin, and several tropic hormones including thyroid stimulating hormone and ACTH.

Fill in the chart: (See following page)

TEST YOURSELF
Match:
11. stimulates water reabsorption
12. stimulates urerine contraction
13. stimulates milk secretion
14. stimulates somatomedin production
15. tropic hormone that stimulates adrenal cortex

a. oxytocin
b. ADH
c. prolactin
d. growth hormone
e. ACTH

Gland	Hormones	Target tissues	Actions	How controlled
Posterior Pituitary	(a)			
	(b)			
	(produced in hypothalamus)			
Anterior Pituitary	(a)			
	(b)			
	(c)			
	(d)			
	(e)			

LO 9: Describe the actions of growth hormone on metabolism, and describe the consequences of hyposecretion and hypersecretion.

Growth hormone stimulates growth mainly by promoting protein synthesis. Hyposecretion during childhood stunts growth; hypersecretion may result in gigantism.

(1) Another name for growth hormone is _____.
(2) What is somatomedin?

325

(3) How does growth hormone affect fat metabolism?
_____

(4) How does growth hormone raise blood sugar level?

(5) Draw a simple diagram to illustrate how the secretion of growth hormone is regulated.

(6) Briefly describe each condition below and give its causes:
   a. acromegaly_____
   b. pituitary dwarfism_____
   _____
   c. gigantism_____

(7) What is the endocrine explanation for the African pygmy?_____
_____

TEST YOURSELF

16. Hyposecretion of growth hormone during childhood may result in   a. acromegaly   b. gigantism   c. pituitary dwarfism   d. cretinism   e. a pygmy

17. Factors that influence secretion of growth hormone include,   a. exercise   b. non-REM sleep   c. emotional attention   d. answers (a),(b) and (c) are correct   d. answers (a) and (b) only are correct

> LO 10: Identify the hormones secreted by the thyroid gland and give their physiologic actions.

The thyroid gland secretes triiodothyronine (T3) and thyroxine (T4) which are known as *the* thyroid hormones. It also produces calcitonin which functions in calcium regulation.

(1) Briefly define or give the significance of:
    a. the iodide pump_____

    b. thyroglobulin_____

    c. PBI_____

    d. thyroid colloid_____

    e. calorigenic effect_____

(2) List the functions of the thyroid hormones:

_____

TEST YOURSELF

18. Thyroid hormones a. increase oxygen consumption by cells b. decrease heat production c. synthesize iodine d. answers (a),(b) and (c) are correct e/ amswers (a) and (b) only are correct

> LO 11: Describe the feedback control mechanisms by which thyroid-hormone secretion is regulated (be able to draw a diagram to illustrate the mechanisms) and describe the thyroid disorders discussed.

When the level of thyroid hormone rises in the blood the pituitary slows its release of thyroid stimulating hormone (TSH). When the level of thyroid hormone falls, TSH secretion increases.

(1) Draw a diagram to illustrate regulation of thyroid hormone secretion:

(2) When and how does the hypothalamus influence thyroid secretion?_____

(3) Briefly describe and give the cause of
   a. cretinism_____
   b. myxedema_____
   c. Grave's disease_____

   d. goiter_____

TEST YOURSELF
Match:

19. characterized by LATS in the blood
20. results from hypothyroidism during childhood
21. hypothyroid condition in adult

a. Grave's disease
b. cretinism
c. myxedema
d. dwarfism

>LO 12: Describe the interrelationships of parathyroid hormone, vitamin D, and calcitonin in regulating calcium levels in the blood and interstitial fluid.

Parathyroid hormone increases calcium levels by stimulating release of calcium from the bones, stimulating calcium reabsorption from the kidney tubules, and activating vitamin D. Vitamin D increases the amount of calcium absorbed from the intestine. When calcium levels become excessive calcitonin is released by the thyroid gland and acts quickly to inhibit calcium reabsorption from bone.

(1) Draw a diagram to show the interrelationship of parathyroid hormone, vitamin D, and calcitonin in regulating calcium levels.

(2) What are the effects of hypoparathyroidism?
_____
_____

(3) What are the effects of hyperparathyroidism?
_____
_____

TEST YOURSELF

22. Parathyroid hormone a. activates vitamin D b. decreases calcium absorption c. inhibits calcium reabsorption in kidneys d. answers (a),(b) and (c) are correct e. answers (b) and (c) only are corrct

> LO 13: Identify the hormones of the islets of Langerhans, and describe the antagonistic relationship of insulin and glucagon in the control of blood-sugar level.

The islets of Langerhans secrete insulin and glucagon. Insulin decreases blood sugar level by facilitating diffusion of glucose into cells. Glucagon increases blood-sugar level by stimulating gluconeogenesis and glycogenolysis.

(1) List 4 specific metabolic actions of insulin:
   1._____
   2._____
   3._____
   4._____

(2) What effect does glucagon have on adipose tissue?
_____

(3) Draw a diagram to illustrate the regulation of insulin and glucagon secretion.

TEST YOURSELF

23. Which of the following are metabolic effects of insulin? a. promotes release of fatty acids from adipose tissue b. stimulates conversion of glycogen to glucose c. stimulates protein synthesis d. answers (a),(b) and (c) are correct e. answers (a) and (b) only are correct

24. When blood-glucose level is high a. alpha cells release glucagon b. beta cells release glucagon c. alpha cells release glucagon and insulin d. alpha cells release insulin e. beta cells release insulin

> *LO 14: Give the disorders associated with malfunction of the islets of the pancreas, describe the physiologic bases of diabetes mellitus and hypoglycemia, and give their clinical and metabolic symptoms.*

The disturbances associated with diabetes can be traced to three effects of insulin deficiency: (1) decreased glucose utilization, (2) increased fat mobilization, (3) increased protein utilization. In hypoglycemia too much insulin is released in response to glucose challenge.

(1) Why is the untreated diabetic thin and emaciated?
_____

(2) Why is ketosis a danger in untreated diabetes?
_____

(3) List two clinical symptoms of diabetes. _____

(4) Give the cause and symptoms of insulin shock.
_____
_____

TEST YOURSELF

25. Which of the following are associated with diabetes: a. ketosis b. acidosis c. glycosuria d. answers (a),(b) and (c) are correct e. answers (a) and (b) only

LO 15: *Describe the role of the adrenal medulla in helping the body to cope with stress, and give the actions of its hormones, epinephrine and norepinephrine.*

Epinephrine and norepinephrine enhance the effects of the sympathetic nervous system . They help the body to respond to stress by increasing heart rate, metabolic rate, and strength of muscle contraction. They also route blood to organs which need more blood in time of stress.

(1) Why is the adrenal medulla sometimes regarded as an extension of the nervous system? _____

(2) What are catecholamines? _____

(3) How do the adrenal medullary hormones affect fatty acid level in the blood? _____

(4) How do they affect blood-sugar level? _____

(5) What stimulates secretion of epinephrine and norepinephrine? _____

TEST YOURSELF

26. Which of the following are actions of the adrenal medullary hormones: a. lower thresholds in RAS b. mobilize fatty acids from adipose tissue c. stimulate glycogenolysis d. answers (a),(b) and (c) are correct e. none of the preceding answers is correct

27. Stimulation of adrenal medullary secretion depends upon a. acetylcholine released by preganglionic neurons of the sympathetic nervous system b. ACTH release by the anterior pituitary c. a releasing factor produced by the hypothalamus d. concentration of parathyroid hormone in the blood e. answers (b) and (c) are correct

LO 16: Describe the actions of mineralocorticoids and glucocorticoids, tell how their secretion is regulated, and give the effects of malfunction.

Aldosterone, the principal mineralocortidoid, increases the rate of tubular reabsorption of sodium in kidneys. Cortisol, the principal glucocorticoid, promotes gluconeogenesis in the liver, thereby raising blood-sugar levels.

(1) How does aldosterone help to regulate blood pressure?_____

(2) What effect does aldosterone have upon potassium?_____

(3) By what mechanisms does cortisol promote gluconeogenesis?_____

(4) Glucocorticoid secretion is regulated by_____ secreted by the pituitary.
ACTH secretion is controlled by_____ _____ hormone from the_____.

(5) Draw a diagram illustrating control of glucocorticoid secretion.

(6) Give the cause of and briefly describe:
a. Cushing's syndrome_____

b. Addison's disease_____
_____

TEST YOURSELF

28. Glucocorticoid secretion increases when a. the hypothalamus releases corticotropin-releasing factor in response to stress b. the anterior pituitary releases ACTH c. sodium concentration decreases in the plasma d. answers (a),(b) and (c) are correct e. answers (a) and (b) only are correct

29. In addison's disease a. the patient loses the ability to cope with stress b. loss of sodium may cause the patient to die of shock c. edema gives the patient a full-moon appearance d. answers (a),(b) and (c) are correct e. answers (a) and (b) only are correct

> LO 17: Describe the body's physiologic responses to stress and the role of the adrenal glands in helping the body to adapt to stress.

In response to stress the adrenal medulla initiates an alarm reaction which prepares the body physiologically to cope with the stress. Secretion of glucocorticoids by the adrenal cortex provides an important backup system by shifting metabolism to high gear so that a steady supply of nutrients is assured to the rapidly metabolizing cells.

(1) What is the GAS?_____
(2) According to Hans Selye the GAS may be divided into 3 stages. Briefly describe the physiologic events of each:
   a. alarm reaction_____
   b. resistance reaction_____
   c. exhaustion stage_____

(3) According to some physiologists chronic stress is harmful. Why?_____

TEST YOURSELF

30. In stress the adrenal cortex provides an important back-up system to the adrenal medulla by   a. stimulating the kidney tubules to release potassium   b. providing the extra nutrients essential to stepped up metabolism   c. lowering blood glucose levels   d. answers (a),(b) and (c) are correct   e. answers (a) and (c) only are correct

> *LO 18: Synthesizing what you have learned about actions of adrenal hormones, growth hormone, pancreatic hormones, and thyroxine, draw a diagram illustrating how blood-sugar level is regulated, or describe this process.*

Fill in the table on the following page:

(1) Tell how thyroid hormone affects blood-glucose level. Be specific. _____

(2) Now, draw a diagram to illustrate some of these interrelationships.

TEST YOURSELF

31. Blood-glucose level is lowered by   a. glucagon   b. insulin   c. growth hormone   d. epinephrine   e. all of the preceding

Hormone	Effect upon glycogen	Effect upon gluconeo-genesis	Effect upon glucose up-take by cells	Overall effect upon blood-sugar level
Epinephrine	stimulates glycogenolysis	stimulates		
Growth hormone		stimulates	decreases uptake	
Insulin	stimulat glycogen formation	inhibits	stimulates uptake	
Glucagon	stimulates	stimulates		

TEST YOURSELF

32. Gluconeogenesis is promoted by a. epinephrine b. glucagon c. insulin d. answers (a), (b) and (c) are correct e. answers (a) and (b) only

33. Glycogen formation is stimulated by a. epinephrine b. glucagon c. insulin d. answers (a), (b) and (c) only are correct e. answers (a) and (b) only are correct

------REVIEW and FILL IN TABLE - NEXT PAGE------
ANSWERS TO TEST YOURSELF QUESTIONS

1.d   2.e   3.d   4.b   5.c   6.a   7.d   8.a   9.b   10.e 11.b 12.a
13.c 14.d 15.e 16.c 17.d 18.a 19.a 20.b 21.c 22.a 23.c
24.e 25.d 26.d 27.a 28.e 29.e 30.b 31.b 32.e 33.c

335

Gland and hormone	Target tissue	Principal actions	How regulated
Thyroid			
Parathyroids			
Islets of Langerhans			
Adrenal Medulla			
Adrenal Cortex			

EFFECTS OF MALFUNCTION

Hormone	Consequence of hyposecretion	Consequence of hypersecretion
Thyroxine		
Parathyroid hormone		
Insulin		
Cortisol		
Aldosterone		

POST TEST

1. (LO 1) Endocrine glands   a. have ducts   b. produce hormones   c. include the sweat glands and gastric glands   d. answers (a),(b) and (c) are correct   e. answers (a) and (b) only are correct

2. (LO 2) Hormones   a. are chemical messengers   b. may be steroids   c. may be proteins or amino acid derivatives   d. answers (a),(b) and (c) are correct   e. answers (a) and (b) only are correct

(LO 3)

3. anatomically associated with the hypothalamus
4. located within the pancreas
5. located anterior to the trachea

a. posterior pituitary
b. adrenal glands
c. thyroid gland
d. islets of Langerhans
e. all of the preceding are correct

6. (LO 4) If we compare hormones to keys, the locks are   a. receptor sites in target tissues   b. releasing factors of the hypothalamus   b. regulating hormones released by the anterior pituitary   d. hormone bound plasma in the blood   e. none of the preceding answers is correct

7. (LO 5) The particular action initiated by cyclic AMP within a target cell depends upon   a. the type of genes within the cell   b. other types of hormones present   c. presence of hypothalamic releasing factors   d. presence of tripic hormones   e. the specific types of enzyme systems present in the cell

8. (LO 6) Which of the following are regulated by negative feedback mechanisms involving the anterior pituitary gland: a. thyroid   b. parathyroid   c. islets of Langerhans   d. answers (a),(b) and (c) are correct   e. answers (a) and (b) only are correct

9. (LO 7) Which of the following glands are regulated (directly or indirectly) by the hypothalamus: a. adrenal cortex b. thyroid c. pituitary d. answers (a),(b) and (c) are correct e. answers (b) and (c) only are correct

10. (LO 7,8) Which of the following are produced by the hypothalamus: a. growth-hormone -inhibiting hormone b. oxytocin c. prolactin d. answers (a) (b) and (c) are correct e. answers (a) and (b) only

(LO 8,9)
11. target tissue is kidney tubules
12. target tissue is uterus
13. promotes protein synthesis
14. tropic hormone

a. oxytocin
b. prolactin
c. ADH
d. ACTH
e. Growth hormone

(LO 9,11, 12, 14, 16)
15. hyposecretion in childhood may lead to cretinism
16. hypersecretion may lead to acromegaly
17. too much causes Cushing's syndrome
18. deficiency related to diabetes mellitus

a. Growth hormone
b. Insulin
c. Thyroxine
d. Glucocorticoids
e. Parathyroid hormone

(LO 9,10,12, 13)
19. stimulates calcium resorption from bone
20. lowers blood calcium levels
21. facilitates glucose diffusion into cells; lowers blood-glucose level
22. stimulates metabolic rate

a. parathyroid hormone
b. calcitonin
c. growth hormone
d. thyroid hormone
e. insulin

(LO 8,12,15,16)
23. maintains sodium balance
24. released by posterior pituitary
25. increases heart rate; raises blood-glucose level
26. activates vitamin D

a. epinephrine
b. aldosterone
c. parathyroid hormone
d. oxytocin
e. prolactin

27. (LO 13) Glucagon is released   a. when insulin level becomes too high   b. by the alpha cells in the islets of Langerhans   c. when blood glucose level falls to about 70 mg/100 ml,   d. answers (a),(b) and (c) are correct   e. answers (b) and (c) only are correct

28. (LO 14)   Insulin deficiency results in   a. decreased protein utilization   b. decreased glucose utilization   c. decreased fat utilization   d. answers (a), (b) and (c) are correct   e. answers (a) and (b) only

29. (LO 17,18) In stress   a. sympathetic nerves signal the adrenal medulla to release greater amounts of its hormones   b. cortisol promotes gluconeogenesis   c. metabolic rate is increased   d. answers (a),(b) and (c) are correct   e. answers (a) and (c) only

30. (LO 18) Which hormones increase blood sugar level?   a. insulin   b. parathyroid hormone   c. glucagon   d. aldosterone   e. three of the preceding answers are correct

ANSWERS TO POST TEST QUESTIONS

1.b	2.d	3.a	4.d	5.c	6.a	7.e	8.a	9.d
10.e	11.c	12.a	13.e	14.d	15.c	16.a	17.d	
18.b	19.a	20.b	21.e	22.d	23.b	24.d	25.a	
26.c	27.e	28.b	29.d	30.c				

18 REPRODUCTION

PRETEST

1. Sperm are manufactured within the tiny, coiled _____ of the testes.
2. Sperm develop from stem cells called _____.
3. How many spermatids are produced from each primary spermatocyte? _____
4. How many chromosomes are present in a normal primary spermatocyte? ____ In a normal spermatid? ____ In a normal mature sperm? _____
5. The _____ at the front tip of the mature sperm is formed from the Golgi.
6. The _____ serves as an air conditioning mechanism for sperm cells.
7. Sperm pass from the epididymis into the _____.
8. The ejaculatory duct opens into the _____.
9. About 60 per cent of the semen volume is produced by the _____.
10. Men with fewer than 40 million sperm per milliliter of semen are usually _____.
11. The two paired cylinders of erectile tissue in the penis are the _____.
12. Erection is a _____ action. During erection the tissues of the corpora cavernosa fill with _____.
13. The male accessory gland through which the ejaculatory duct passes is the _____.
14. Testosterone is produced primarily by the _____ cells of the testes.
15. In the male, FSH acts to promote _____.
16. An ovum and its granulosa cells constitute a _____.

17. In oogenesis the second meiotic division does not occur until after _____.
18. The _____ is a temporary endocrine structure within the ovary. It produces _____.
19. The part of the uterus that projects into the vagina is the _____.
20. Fertilization normally occurs in the upper third of the _____.
21. The numerous transverse ridges within the vaginal wall are called _____.
22. The two openings within the vestibule are the _____ orifice, anteriorly, and the _____ orifice posteriorly.
23. The _____ is a small body of erectile tissue in the female that is homologous to the male glans penis.
24. The secretory cells of the breasts are arranged in grapelike clusters called _____.
25. When the infant suckles at the breast a reflex action results in release of the hormones _____ and _____ from the pituitary gland.
26. Mammography is a soft-tissue radiologic study of the _____.
27. The most common site of cancer in women is _____.
28. _____ begins on the first day of the menstrual cycle; _____ occurs about 14 days before the next cycle begins.
29. In the female, growth of sec organs at puberty is stimulated mainly by the hormone _____.
30. Development of the corpus luteum is stimulated by _____.
31. _____ occurs at about age 50 marking the end of a woman's reproductive ability.
32. In the male the first response to sexual excitement is _____; in the female it is _____.
33. Sexual intercourse is most likely to result in pregnancy wh it occurs around day _____, of a typical menstrual cycle.
34. When a Y sperm fertilizes the ovum, the offspring will be _____.
35. Oral contraceptives prevent pregnancy by preventing _____.
36. In vasectomy each _____ _____ is severed and tied.

37. _____ abortions are performed in order to preserve the health of the mother or when there is significant risk of a malformed child.
38. The first symptom of syphilis is a _____ _____.
39. Genital _____ is an increasingly common sexually transmitted disease caused by a virus.
40. The sexually transmitted disease _____ is the second most common communicable disease.

ANSWERS: 1. seminiferous tubules (LO 1) 2. spermatogonia (LO 2)   3. four (LO 2,3)   4. 46; 23; (LO 2,3)
5. acrosome (LO 4)   6. scrotum (LO 5)   7. vas deferens (LO 6)   8. urethra (LO 6)   9. seminal vesicles (LO 7, 8)   10. sterile (LO 8)   11. corpora cavernosa (LO 9)
12. reflex, blood (LO 10)   13. prostate (LO 7,11)
14. interstitial (or Leydig) (LO 12)   15. spermatogenesis (LO 12)   16. follicle (LO 13)   17. fertilization (LO 13)   18. corpus luteum, progesterone (also estrogen) (LO 14)   19. cervix (LO 15)   20 uterine tube (LO 15)   21. rugae (LO 16)   22. urethral, vaginal (LO 16)   23. clitoris (LO 16)   24. alveoli (LO 17)
25. prolactin, oxytocin (LO 18)   26. breast (LO 19)
27. the breast (LO 19)   28. menstruation (bleeding), ovulation (LO 20)   29. estrogen (LO 21)   30. LH (luteinizing hormone) (LO 21)   31. menopause (LO 22)
32. penile erection, vaginal lubrication (LO 23)
33. 14 (LO 24)   34. male (LO 26)   35. ovulation (LO 26)
36. vas deferens (LO 27) 37. therapeutic (LO 28)
38. primary chancre (LO 29)   39. herpes (LO 29)
40. gonorrhea (LO 29)

> LO 1: Describe the internal structure of the testes,

Each testis is covered by a tunica albuginea and is partially divided by septa; the seminifierous tubules lie in the lobes between the septa.

(1) Briefly describe
   a. tunica albuginea _____
   b. mediastinum (of testis)
(2) Label the diagram of the testis on the following page:

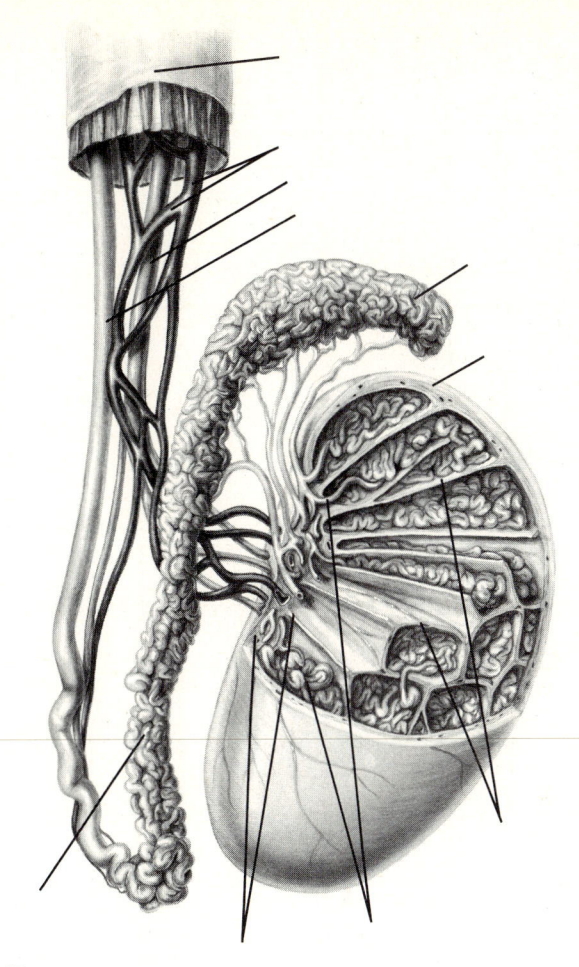

TEST YOURSELF
Match:
1. tiny coiled tubules; site of spermatogenesis
2. white sheath of connective tissue covering testis

a. septum
b. tunica albuginea
c. mediastinum
d. seminiferous tubules
e. rugae

   LO 2: Describe spermatogenesis, and draw a diagram to illustrate the process.

Some spermatogonia differentiate to become primary spermatocytes which undergo meiosis to form secondary spermatocytes, then spermatids. The four spermatids

343

derived from each primary spermatocyte differentiate to form mature sperm cells.

(1) How are developing sperm cells arranged within a seminiferous tubule? Draw a diagram to illustrate.

(2) Draw a diagram to illustrate spermatogenesis.

TEST YOURSELF

3. The developing sperm cells that go through the first meiotic division are   a. primary spermatocytes   b. spermatids   c. spermatozoa   d. spermatogonia   e. secondary spermatocytes

4. The cells found next to the basement membrane of a seminiferous tubule are the   a. spermatids   b. spermatozoa   c. spermatogonia   d. primary spermatocytes   e. secondary sperm atocytes

>LO 3: *Give the essential features of meiosis, and contrast meiosis with mitosis.*

In meiosis each primary spermatocyte or oocyte gives rise to four cells, each with the haploid number of chromosomes. Homologous chromosomes come together during the prophase of the first meiotic division forming tetrads, and crossing over occurs.

Fill in the table on the following page:

	Mitosis	Meiosis
a. number of cells produced		
b. number of divisions		
c. homologous chromosomes come together (synapsis occurs)?		
d. crossing over occurs?		
e. number of chromosomes in each daughter cell produced		
f. When do chromatids (duplicated chromosomes) separate?		

TEST YOURSELF
Match:

5. crossing over occurs
6. synapsis occurs
7. four daughter cells produced
8. haploid daughter cells produced
9. daughter cells have identical chromosome composition as mother cell

a. mitosis only
b. meiosis only
c. both mitosis and meiosis
d. neither

LO 4: Draw and label a diagram of a mature sperm cell, and give the function for each of its unique structures (e.g. acrosome and tail).

A mature sperm cell consists of a head, midpiece, and a tail which is used for locomotion. The head contains the nucleus and at its front tip the acrosome.

(1) What is the acrosome made up of?
(2) What is the function of the acrosome?
(3) What organelles are concentrated in the midpiece?
(4) Draw and label a diagram of a sperm cell. (If necessary consult Fig. 18-7 in your textbook)

TEST YOURSELF

10. Mitochondria of the mature sperm are concentrated within the   a. acrosome   b. nucleus   c. tail   d. head   e. midpiece

11. The acrosome contains   a. mitochondria   b. DNA   c. ATP   d. enzymes thought to aid in fertilization   e. carbohydrate needed for locomotion

> LO5: Describe the descent of the testes into the scrotum and give the function of the scrotum.

Before birth the testes descend through the inguinal canals into the scrotum. The scrotum maintains the sperm at a temperature about $2°$ C cooler than body temperature.

(1) As the testes descend into the scrotal sac they pull their arteries,_____,_____, and _____ along after them. These structures encased by the _____ muscle and by layers of fascia constitute the _____.

(2) What is an inguinal hernia?_____

(3) What are the physiologic mechanisms by which a low temperature is maintained within the scrotum?
_____
_____

(4) What is the function of the dartos muscles?
_____

(5) What type of clinical problem results from untreated cryptorchidism?_____
_____

TEST YOURSELF

12. In cold weather   a. the dartos muscles contract   b. the spermatic cord contracts   c. the testes are positioned close to the abdominal wall   d. answers (a),(b) and (c) are correct   e. none of the preceding answers is correct

13. Cryptorchism   a. can lead to sterility   b. may cause feminization of a male   c. is a frequent cause of inguinal hernia   d. answers (a),(b) and (c) are correct   e. none of the preceding answers is correct

> LO 6: Trace the passage of sperm from the seminiferous tubules through the conducting tubes, describing changes that may occur along the way.

(1) Study the Fig. on p. 351   in your text book) Number each structure through which sperm pass.
(2) Sperm cells pass from the seminiferous tubules through passageways in the _____ and through small tubules which exit from the testes into the epididymus. There they are _____ and continue their maturation. From the epididymus sperm pass into the _____ where further _____ may take place. During ejaculation sperm pass from the vas deferens into the _____ duct and then out through the _____ which passes through the _____.
(3) Which of the conducting tubes is only a single tube (i.e. unpaired)? _____

TEST YOURSELF

14. From the vas deferens sperm pass directly into the   a. mediastinum   b. epididymus   c. ejaculatory duct   d. urethra   e. penis

15. Which of the following are unpaired (i.e. single) structures: a. epididymides   b. vas deferentia   c. ejaculatory duct   d. testes   e. urethra

> LO 7: Name the reproductive accessory glands in the male (or label on a diagram), and give their functions.

Three types of accessory glands contribute to the semen. The _seminal vesicles_ secrete a mucoid fluid rich in fructose which nourishes the sperm and prostaglandins which may help in sperm transport. The _prostate_ gland secretes a thin, milky alkaline fluid important in neutralizing the acidity of other fluids in the semen.

The *bulbourethral glands* release a few drops of alkaline fluid which may be helpful in neutralizing the urethra and in lubrication.

(1) Locate and label the accessory glands on Fig.p. 351
(2) Why is it important that the semen be neutralized by the secretion of the prostate?_____
_____

(3) What is the function of Cowper's glands? What is another name for these glands?_____

(4) How does disorder of the prostate sometimes affect urination?_____
_____

TEST YOURSELF

16. Semen is neutralized by the alkaline secretion of the a. seminal vesicles   b. bulbourethral glands   c. Cowper's glands   d. glands in the wall of the vas deferens   e. prostate gland

17. When the acidity of the semen is neutralized   a. prostatic cancer may develop   b. sperm become active   c. erection is stimulated   d. the male may become temporarily sterile   e. answers (a) and (d) are correct

> LO 8: Give the composition of semen, and describe reasons for male sterility.

Semen from one ejaculation consists of about 400 million sperm cells suspended in about 3.5 ml of fluid from the accessory glands and walls of the ducts. Sperm cells are so tiny that they account for very little of the semen volume.

(1) Give the relative volume contribution and briefly describe the secretions of
 a. seminal vesicles_____

 b. prostate_____

 c. bulbourethral glands_____

(2) Give two reasons for male sterility_____
_____

TEST YOURSELF

18. Male sterility may result from: a. semen containing fewer than 40 million sperm per ml b. large numbers of abnormal sperm in the semen c. fever or infection d. answers (a),(b) and (c) are correct e. answers (a) and (c) only are correct

> LO 9: Describe(and be able to label on a diagram) the internal and external structures of the penis and give their functions.

The shaft of the penis enlarges distally to form the glans which may be partially covered by the prepuce. Internally the penis consists of three cylinders of erectile tissue.

(1) Describe the erectile tissue of the corpora cavernosa_____

(2) What part of the penis is removed furing circumcision?_____
(3) Label the structures of the penis illustrated in the Fig. on p. 351.

TEST YOURSELF

19. The erectile tissue of the penis is composed mainly of a. muscle b. fibrous connective tissue c. large venous sinusoids d. arterioles e. bone and muscle

> LO 10: Describe the physiologic bases for erection and ejaculation.

Erection and ejaculation are spinal reflexes. Erection occurs when arteries of the penis dilate, allowing more blood to enter the erectile tissue than can leave. Ejaculation consists of emission, when the accessory glands eject their fluids into the urethra, followed by ejection of the semen from the penis.

(1) What types of stimuli initiate the erection reflex? _____

(2) Which nerves convey the message to become erect from the spinal cord to the penis? _____

(3) How does the penis become flacid once again? _____

(4) What is impotence? _____
_____

TEST YOURSELF

20. Ejaculation a. is a spinal reflex b. is stimulated by sympathetic nerves from the lumbar region of the spinal cord c. involves emission and actual ejection of semen d. answers (a),(b) and (c) are correct e. answers (a) and (c) only are correct

> LO 11: Label diagrams of internal and external male reproductive organs, and describe their structure and functions.

Male reproductive structures consist of the testes where sperm production takes place and the scrotum which houses the testes; the system of ducts through which sperm are transported; the accessory glands which produce the semen; and the penis which delivers the sperm into the female reproductive system.

Label the diagram on the following page.

TEST YOURSELF
Match:

21. produce sperm
22. accessory glands
23. conduct semen to ejaculatory ducts

a. scrotum
b. vas deferens
c. seminal vesicles
d. penis
e. testes

> LO 12: Describe the actions of the male gonadotropic hormones and of testosterone, and tell how they are controlled.

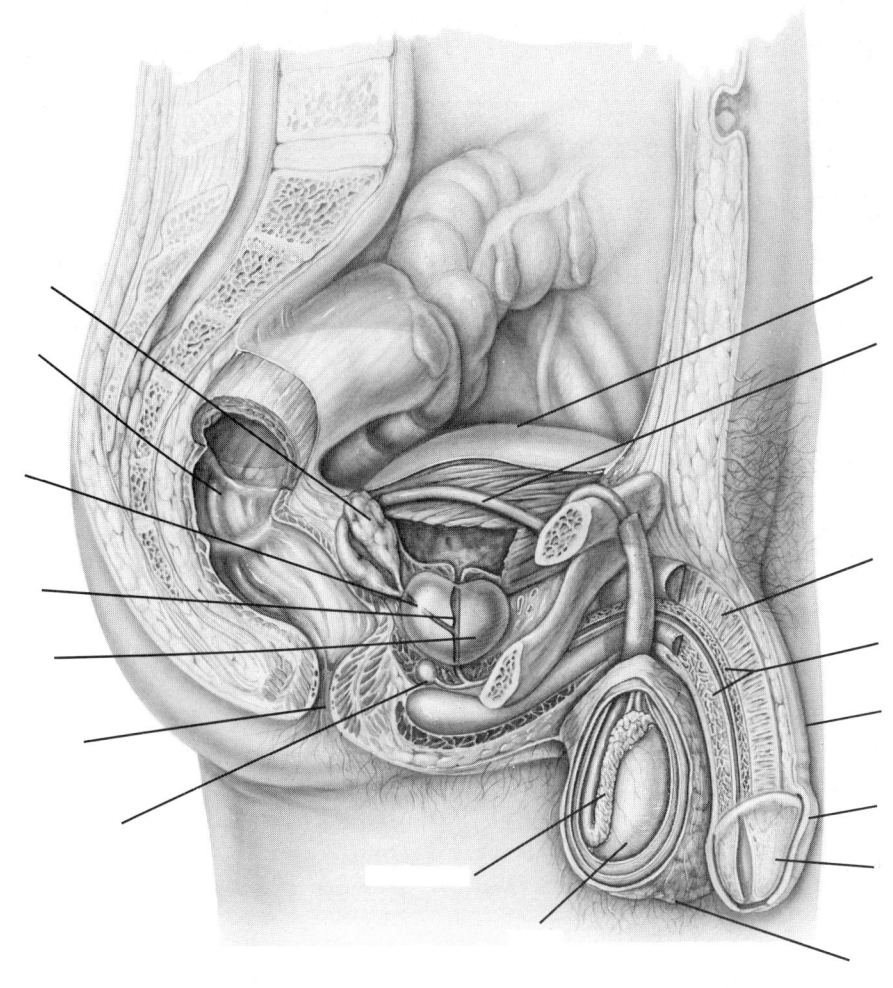

FSH promotes spermatogenesis whereas LH stimulates the interstitial cells to secrete testosterone. Testosterone stimulates growth of the primary sex organs and the development of the secondary sex characteristics at puberty. It also stimulates general body growth.

(1) What type of chemical compound is testosterone?

(2) Give the actions of testosterone:
    a. before birth _____
    b. at puberty _____
    c. in the adult _____

(3) On a separate sheet of paper draw a diagram to illustrate the regulation of male hormone secretion.

TEST YOURSELF
Match:
24. stimulates growth of beard
25. stimulates spermatogenesis
26. responsible for growth spurt
27. stimulates testosterone production

a. testosterone
b. FSH
c. LH
d. estrogen
e. none of the above

28. If a male is castrated before puberty a. masculinity is not affected b. secondary sex characteristics may be affected, but not sex drive or performance c. the male will grow to be slightly taller than normal d. the epiphyses close sooner than normal e. two of the preceding answers are correct

> LO 13: Trace the development of an ovum in time and space from oogonium through fertilization.

Oogonia develop in the embryo and give rise to primary oocytes. Each primary oocyte lies within a layer of granulosa cells and is part of a follicle. After puberty a few follicles begin to develop each month under the influence of FSH. The first meiotic division yields a secondary oocyte and smaller polar body. The second meiotic division takes place only after fertilization occurs. At ovulation the secondary oocyte is ejected into the abdominal cavity and then passes into the uterine tube where it is either fertilized or deteriorates.

(1) Identify and briefly describe
    a. zona pellucida _____
    b. theca _____
    c. graafian follicles _____
    d. atretic follicles _____
(2) On a separate sheet of paper draw your own series of diagrams illustrating the development of an ovum.
(3) Label the diagram, on the following page, illustrating the microscopic structure of the ovary.

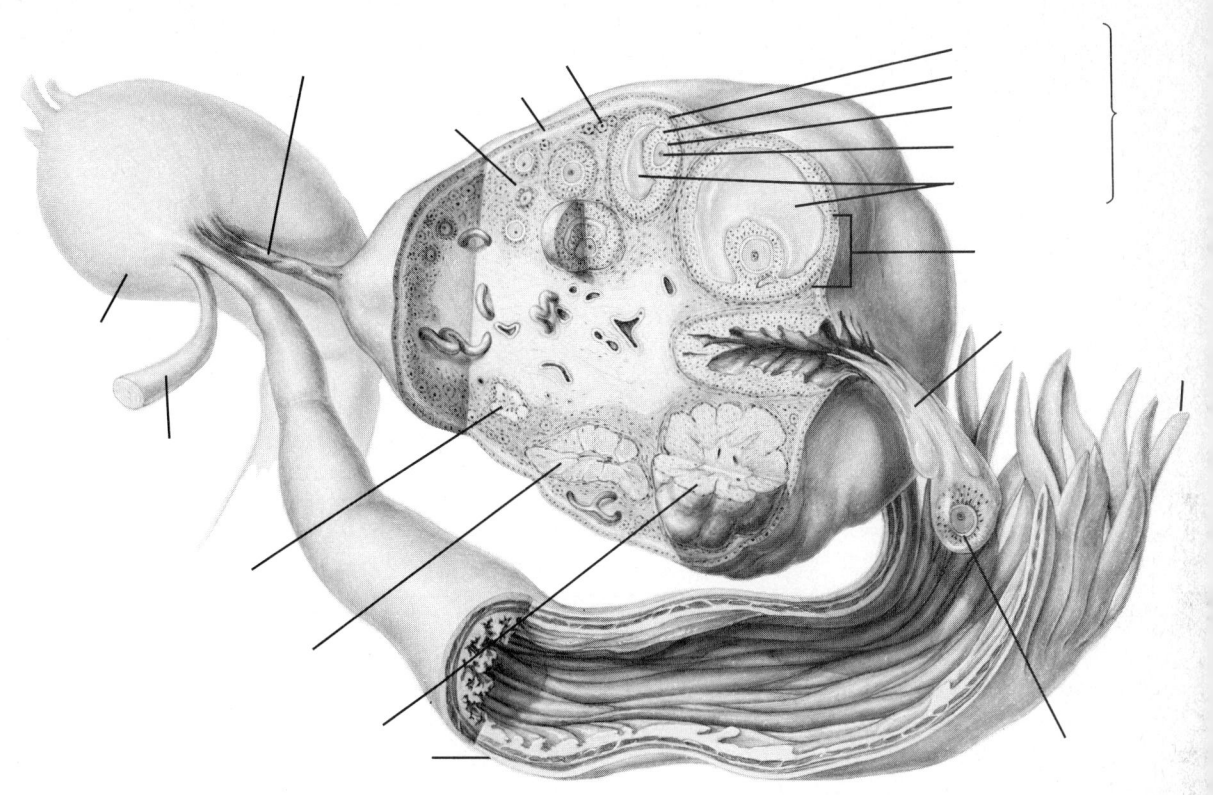

TEST YOURSELF
Match:
29. No new ones arise after birth
30. Mature follicle
31. The smaller of the two cells derived from the first meiotic division
32. Does not divide until after fertilization

a. Graafian follicle
b. atretic follicle
c. oogonia
d. polar body
e. secondary oocyte

33. The cells that secrete estrogen are a. found within the zona pellucida b. the oogonia c. the atretic follicles d. the primary oocytes e. found within the theca

> LO 14: Trace the development of a corpus luteum and give its function.

Under the influence of LH the part of the follicle left after ovulation develops into a corpus luteum. The corpus luteum is a temporary endocrine structure which secretes estrogen and progesterone. If fertilization does not occur the corpus luteum begins to degenerate after about ten days. Should fertilization occur the corpus luteum continues to secrete progesterone for several months.

(1) What is a corpus albicans?_____
(2) What are luteal cells?_____
(3) Which hormones are produced by the corpus luteum?
_____

TEST YOURSELF

34. A corpus luteum   a. develops from the ruptured follicle   b. develops under the influence of LH   c. eventually develops into a corpus albicans   d. answers (a),(b) and (c) are correct   e. answers (a) and (b) only are correct

> LO 15: Describe and give the functions of each internal organ of the female reproductive system, and be able to draw and label diagrams of these structures.

Internal organs of the female reproductive system include (1) the <u>ovaries</u> which produce ova and the hormones estrogen and progesterone, (2) the <u>uterine</u> (fallopian) <u>tubes</u> which function in ovum transport and as the site of fertilization, (3) the <u>uterus</u>, incubator for the developing embryo, (4) the <u>vagina</u> which functions as a coital organ, as part of the birth canal, and as an exit for menstrual discharge.

(1) Label the diagrams on the following <u>two</u> pages.

(2) What is (are) the
   a. endometrium_____

   b. myometrium_____

   c. rugae of the vagina_____
   _____

TEST YOURSELF

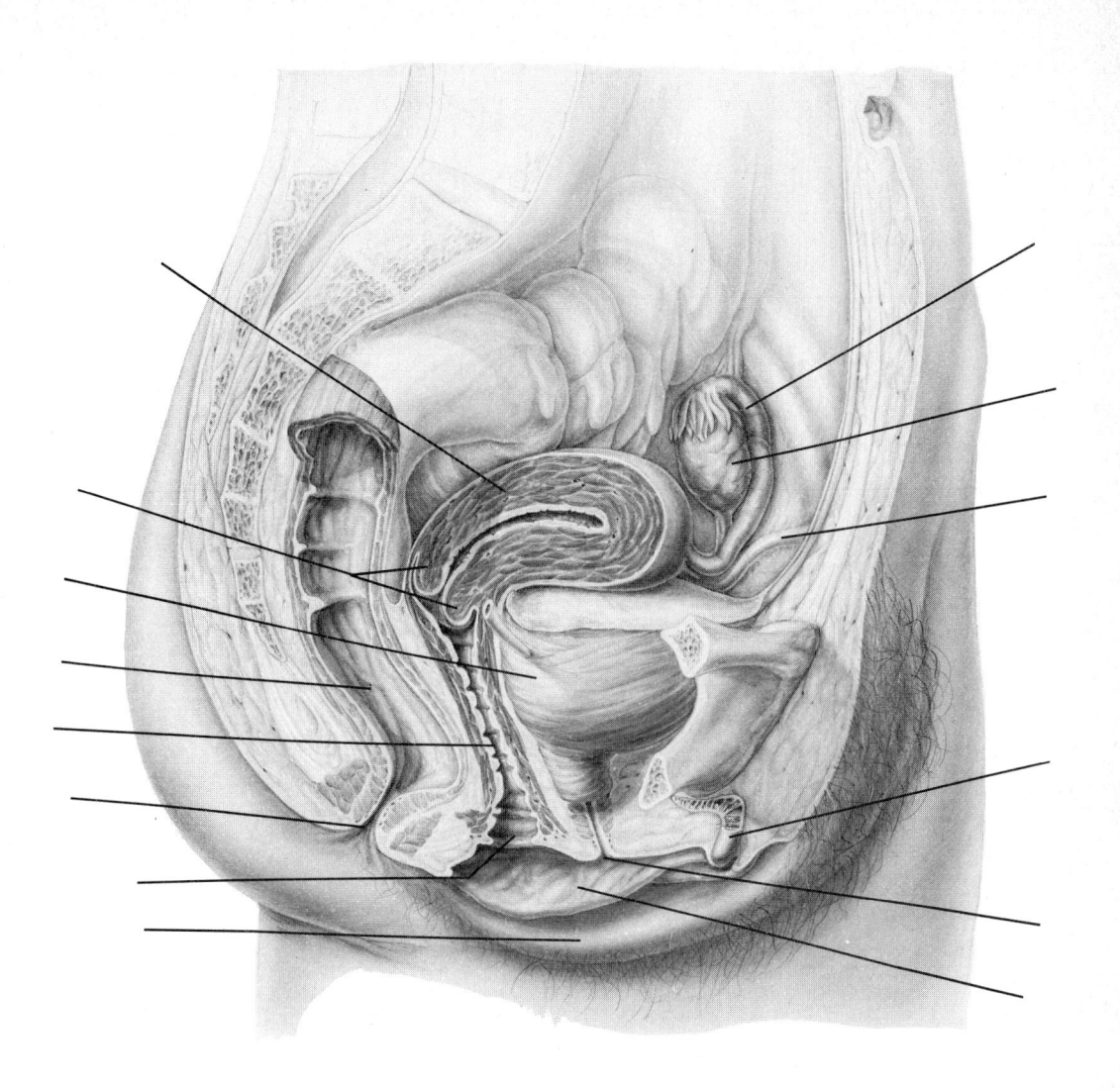

Match:
35. ova production
36. site of fertilization
37. incubator for developing embryo
38. estrogen production

a. uterine tube
b. uterus
c. vagina
d. ovary
e. rugae

LO 16: Label diagrams of, describe, and give the functions of the structures of the vulva.

The vulva consists of mons pubis, labia majora, labia minora, clitoris, and vestibule of the vagina.

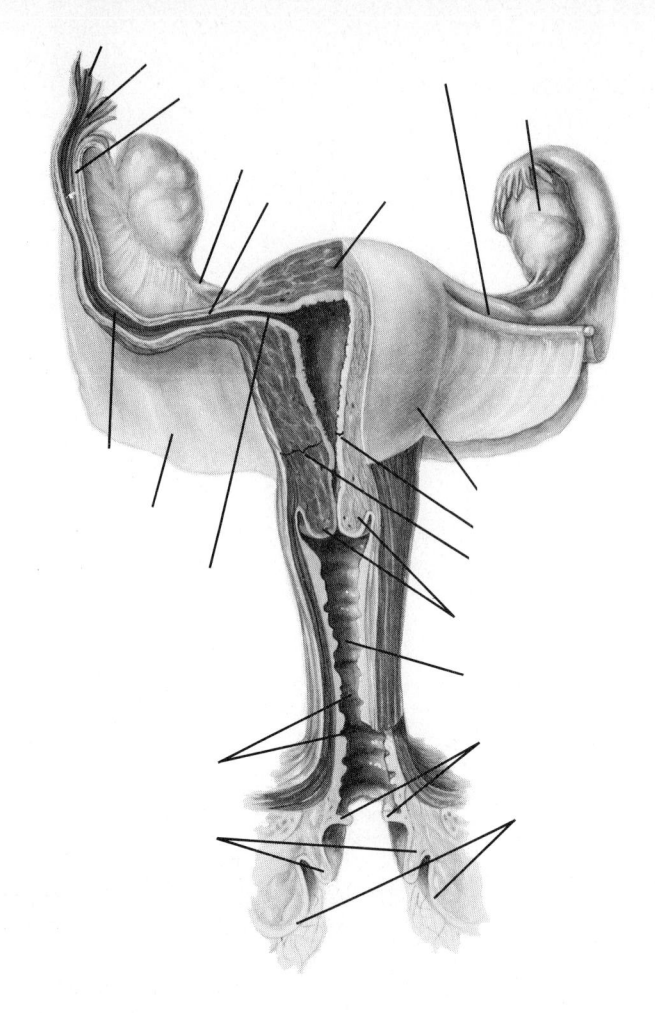

(1) What is the function of the clitoris? _____
_____
(2) Briefly identify_____
   a. hymen_____
   b. Bartholin's glands_____
   c. bulbs of the vestibule_____
   d. perineum_____
TEST YOURSELF
Match:
39. help protect genital structures   a. perineum
40. small body of erectile tissue   b. labia majora
    hemologous to glans penis   c. clitoris
41. region between vaginal orifice and d. mons pubis
    anus   e. hymen

Label the diagram:

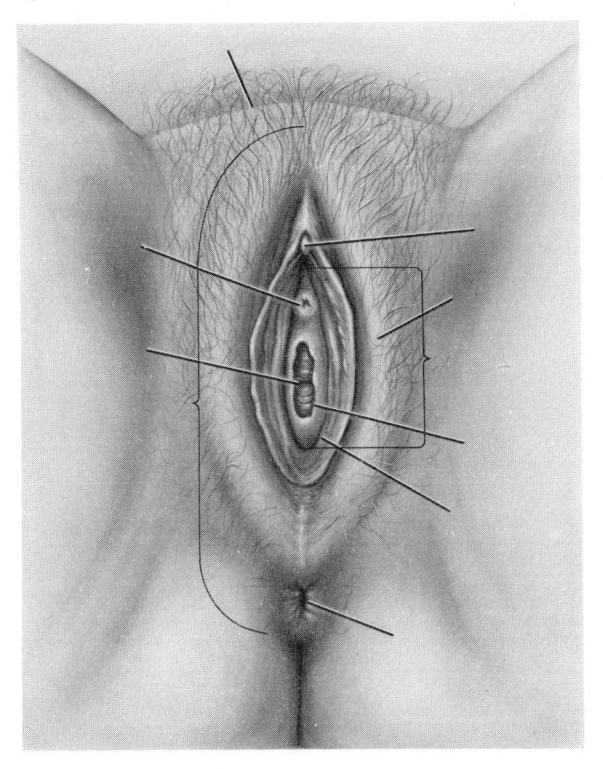

LO 17: Describe the development and structure
of the breasts, relating these to function.

Secretory cells of the breasts are arranged in alveoli.
Ducts from alveolar clusters unite to form a single
duct from each lobe, and these main ducts open on
the nipple surface.

(1) The breasts overlie the _____ muscle and are
    attached to them by fascia, Ligaments of _____
    firmly connect the breasts to the skin.
(2) Describe the
    a. nipple _____
    b. areola _____

TEST YOURSELF

42. The pinkish area surrounding the nipple is the
    a. Cooper ligament   b. alveolus   c. ampulla   d. fascia   e. areola

LO 18: List the hormones that affect breast development and lactation, and give their functions.

Estrogen and progesterone stimulate breast development during puberty and again during pregnancy. When an infant suckles at the breast oxytocin and prolactin are released from the pituitary gland. Prolactin stimulates milk production and oxytocin promotes release of milk into the ducts.

Complete the table

Hormone	Function (with respect to the breasts)
Estrogen	
Progesterone	
Prolactin	
Oxytocin	

TEST YOURSELF
Match
43. release of milk from glands
44. milk production
45. promotes water retention

a. estrogen
b. progesterone
c. oxytocin
d. prolactin
e. FSH

LO 19: Outline detection methods, course, and treatment of breast and cervical cancer.

Breast cancer may be identified by self-examination or by routine examinations by a physician combined with such diagnostic techniques as mammography. Cervical cancer may be detected by routine Pap smear; incidence appears to be highest in those who have frequent intercourse with many partners.

The_____ are the most common site of cancer in women.
Most cervical cancer is of the_____ type.

TEST YOURSELF

46. In the Pap test a a. few cells are scraped from the cervix b. cells are examined microscopically c. the lesion is treated with radiation d. answers (a),(b) and (c) are correct e. answers (a) and (b) only are correct

> *LO 20: Describe the principal events of the menstrual cycle, correlating pituitary, hypothalamic, ovarian, and uterine responses and interactions.*

The first day of menstruation is counted as the first day of the menstrual cycle. FSH stimulates follicle growth during the preovulatory phase of the cycle. As the follicles grow they release estrogen which stimulates proliferation of the endometrium. At about the 14th day of a "typical" cycle LH and FSH stimulate ovulation. In the postovulatory phase progesterone, secreted by the corpus luteum, is the dominant hormone. It stimulates the glands in the endometrium to secrete a nutritive fluid. When the corpus luteum begins to degenerate estrogen and progesterone levels fall and the uterine lining begins to slough off once again.

(1) Indicate on the line diagram the important events of the menstrual cycle and indicate which hormones are dominant at different times of the cycle.

Day 1   14   28

(2) At which time of a typical 28 day menstrual cycle are each of the following most prominent?
    a. FSH _____
    b. Progesterone _____
    c. The corpus luteum_____
    d. Sudden drop in estrogen progesterone levels
_____

TEST YOURSELF

Match (for a typical cycle)

47. first day of menstruation
48. progesterone dominant

a. day 14
b. day 1
c. first half of cycle
d. second half of cycle
e. day 5

> LO 21: List the principal actions of the gonadotropic and ovarian hormones, and diagram feedback mechanisms involved in their regulation.

FSH stimulates development of follicles at the beginning of each menstrual cycle. LH stimulates development of a corpus luteum after ovulation. Together the two hormones stimulate ovulation. Estrogen is responsible for growth of sex organs at puberty and for development and maintenance of secondary sex characteristics. Each month it stimulates the endometrium to thicken. Progesterone also acts upon the endometrium stimulating development of glands which secrete glycogen.

(1) Fill in the table

Hormone	Principal target tissue	Principal actions
FSH		
LH		
Estrogen		
Progesterone		

(2) On a separate sheet of paper draw a diagram to show how the levels of these hormones are regulated.

TEST YOURSELF
Match
49. stimulates development of follicles
50. stimulates development of the endometrium during the first half of the cycle

a. progesterone
b. LH
c. FSH
d. prolactin
e. estrogen

> LO 22: Describe the physiologic basis and
> reactions of menopause.

At menopause the ovaries become less responsive to FSH and LH, and they secrete less estrogen and progesterone. As a result the ovaries and other reproductive structures begin to atrophy.

What is the hormonal basis of menopause?_____
_____
_____

TEST YOURSELF

51. At menopause a. the menstrual cycle becomes irregular and eventually halts b. less estrogen and progesterone is produced c. ovaries begin to degenerate d. answers (a),(b) and (c) are correct e. none of the preceding answers is correct

> LO 23: Describe the physiologic changes that
> take place during sexual response and compare male and female response.

Vasocongestion and myotonia are two basic general physiologic responses to sexual stimulation. The sexual response cycle may be divided into four phases: excitement, plateau, orgasm, and resolution.

(1) What is meant by
   a. vasocongestion_____
   _____
   b. myotonia_____
(2) Briefly describe and compare:

	in male	in female
a. excitement phase	_____	_____
b. plateau	_____	_____
c. orgasm	_____	_____
d. resolution	_____	_____

> LO 24: Describe the process of conception, and identify the time of the menstrual cycle at which sexual intercourse is most likely to result in conception.

Once released into the vagina sperm must pass through the cervix, uterus, and up the uterine tube in order to reach the ovum. Enzymes from the acrosome help to break down barriers around the egg. Sexual intercourse is most likely to result in pregnancy when it occurs about the time of ovulation.

(1) Why are the presence of millions of sperm in the semen necessary in order for a male to be fertile? _____

(2) Briefly describe fertilization. _____
_____
_____

TEST YOURSELF

53. During most of the menstrual cycle   a. the vagina is so acidic that it is a hostile place for sperm cells   b. a thick plug of mucus blocks the entrance into the uterus   c. cervical mucus rich in glycogen nourishes entering sperm   d. answers (a) and (b) are correct   e. none of the preceding answers is correct

> LO 25: Explain specifically how the sex of the offspring is determined.

The male sex cell determines the sex of the offspring and this determination is made at the moment of conception.

(1) If an X-bearing sperm fertilizes the ovum the offspring will be _____.
(2) If a Y-bearing sperm fertilizes the ovum the offspring will be _____.

TEST YOURSELF
54. In order to produce a male child   a. the ovum must be released by the right ovary   b. the ovum must be

fertilized by a Y sperm   c. the ovum must be fertilized by an X sperm   d. the semen must contain a higher proportion of Y than X sperm   e. answers (a) and (b) are correct

>LO 26: Describe the mode of action of the methods of contraception described and give advantages and disadvantages of each.

(1) Give the mode of action of:
a. oral contraceptives_____

b. IUD_____

c. Spermicides_____

d. Condom_____

e. Diaphragm_____

f. Rhythm_____

TEST YOURSELF
Match:
55. blocks entry of sperm into cervix
56. chemically kills sperm
57. prvvents ovulation

a. contraceptive diaphragm
b. IUD
c. oral contraceptive
d. foam, jelly
e. rhythm

>LO 27: Compare male and female sterilization procedures.

Vasectomy is a relatively simple surgical procedure in which the vas deferentia are cut and tied so that sperm can no longer leave the scrotum. In tubal ligation, the uterine tubes are severed and tied so that no ovum can traverse them.

(1) Why is tubal ligation generally considered more complicated surgery than vasectomy?_____
_____

(2) How is masculinity affected by vasectomy?_____

TEST YOURSELF

58. In vasectomy   a. the vas deferentia are severed so that sperm can no longer leave the scrotum   b. the testes are removed   c. testosterone production is permanently halted   d. answers (a),(b) and (c) are correct   e. answers (a) and (c) only are correct

> *LO 28: Describe the common methods of inducing abortion and discuss the safety of abortion.*

First trimester abortions are performed using a suction method; later abortions use saline injections. Abortions are safest when performed during the first trimester.

(1) Briefly define
   a. spontaneous abortion_____

   b. therapeutic abortion_____

(2) Compare the safety of abortions with the safety of oral contraception and with childbirth._____

TEST YOURSELF

59. A spontaneous abortion   a. is performed using prostaglandins   b. always takes place during the first trimester   c. is a miscarriage   d. answers (a) and (b) are correct   e. none of the preceding answers is correct.

> *LO 29: Describe the sexually transmitted diseases syphilis, gonorrhea, and genital herpes with respect to their mode of infection, course, clinical symptoms and diagnosis, and treatment.*

Fill in the chart on the following page.

	Mode of infection	Principal symptoms	Course of disease
Gonorrhea	_____	_____	_____
	_____	_____	_____
	_____	_____	_____
Syphilis	_____	_____	_____
	_____	_____	_____
	_____	_____	_____
Genital herpes	_____	_____	_____

TEST YOURSELF

60. Gonorrhea   a. is caused by a bacterium   b. is transmitted by sexual contact   c. causes no symptoms in about 80 percent of infected women   d. answers (a),(b) and (c) are correct   e. answers (a) and (b) only.

61. In syphilis   a. a primary chancre is generally the first symptom   b. a secondary stage occurs sometimes marked by a characteristic rash   c. if untreated the disease enters a latent stage which may last for many years   d. answers (a),(b) and (c) are correct   e. answers (a) and (b) only

ANSWERS TO TEST YOURSELF QUESTIONS

1.d	2.b	3.a	4.c	5.b	6.b	7.b	8.b	9.a
10.e	11.d	12.d	13.a	14.c	15.e	16.e	17.b	
18.d	19.c	20.d	21.e	22.c	23.b	24.a	25.b	
26.a	27.c	28.c	29.c	30.a	31.d	32.e	33.e	
34.d	35.d	36.a	37.b	38.d	39.b	40.c	41.a	
42.e	43.c	44.d	45.b	46.e	47.b	48.d	49.c	
50.e	51.d	52.a	53.d	54.b	55.a	56.d	57.c	
58.a	59.c	60.d	61.d					

POST TEST

1. (LO 1) The seminiferous tubules   a. lie within the lobes of the testes   b. are the site of sperm production   c. are where meiosis takes place   d. answers (a),(b) and (c) are correct   e. answers (a) and (b) only are correct

2. (LO 2) The stem cells for sperm are   a. spermatozoa   b. spermatogonia   c. oogonia   d. primary spermatocytes   e. spermatids

3. (LO 3) In meiosis   a. homologous chromosomes come together   b. crossing over takes place   c. four daughter cells are produced   d. each daughter cell is haploid   e. all of the preceding answers are correct

4. (LO 4) At the front tip of a mature sperm cell you could identify the   a. acrosome   b. mitochondria   c. nucleus   d. ATP   e. tail

5. (LO 5) The testes   a. develop in the abdominal cavity of the embryo   b. descend into the scrotum before birth   c. require a higher than body temperature for sperm production   d. answers (a),(b) and (c) are correct   e. answers (a) and (b) only.

6. (LO 6,11) From the epididymus sperm pass directly into the   a. ejaculatory duct   b. urethra   c. vas deferens   d. seminiferous tubule   e. mediastinum

7. (LO 7, 11, 8) The secretion of the seminal vesicles   a. is thin and alkaline   b. neutralizes the acidity of the other fluids contributing to the semen   c. contains fructose and other nutrients as well as prostaglandins   d. answers (a),(b) and (c) are correct   e. answers (a) and (b) only.

8. (LO 8, 10) John's semen is clinically analyzed and is found to contain about 3 million sperm per milliliter of semen. John is probably   a. impotent   b. sterile   c. castrated   d. oversexed   e. a eunuch

9. (LO 9) Which of the following is (are) removed during circumcision: a. glans   b. corpora cavernosa

c. prepuce   d. answers (a),(b) and (c) are correct
e. answers (a) and (b) only.

10. (LO 10) Erection   a. is a reflex response to sexual excitement   b. occurs when more blood enters the penis than can leave   c. involves emission and actualy ejection of semen   d. answers (a),(b) and (c) are correct   e. answers (a) and (b) only

(LO 11, 7) refers to questions 11-14.
11. passes through penis
12. maintains cool temperature for sperm
13. produce(s) alkaline secretion
14. site of testosterone production

a. scrotum
b. testes
c. urethra
d. prostate
e. vas deferens

15. (LO 12) In the male LH secretion is inhibited by
a. high levels of testosterone in the blood
b. low levels of FSH   c. low levels of estrogen
d. answers (a),(b) and (c) are correct   e. none of the preceding answers is correct.

16. (LO 12) Testosterone acts to   a. stimulate descent of testes before birth   b. stimulate growth spurt at puberty   c. maintain secondary male sex characteristics   d. answers (a),(b) and (c) are correct
e. answers (b) and (c) only.

17. (LO 11) Which of the following may be part of a follicle   a. oocyte   b. granulosa cells   c. theca
d. answers (a),(b) and (c) are correct   e. answers (a) and (c) only.

18. (LO 13, 14) After ovulation   a. no part of the ejected follicle remains in the ovary   b. some remains of the follicle stay in the ovary as an atretic follicle   c. remains of the follicle within the ovary develop into a corpus luteum   d. a corpus albicans forms within 24 hours   e. two of the preceding answers are correct.

19. (LO 14) The corpus luteum   a. releases estrogen
b. releases progesterone   c. formed each month remains as an endocrine structure until menopause
d. answers (a),(b) and (c) are correct   e. answers (a) and (b) only.

(LO 15) refers to questions 20-24
20. free end of uterine tube
21. portion of uterus that projects into vagina
22. folds in wall of vagina
23. muscle layer of uterus
24. longest portion of uterine tube

a. ampulla
b. myometrium
c. infundibulum
d. cervix
e. rugae

25. (LO 16) The mons pubis   a. is part of the perineum   b. is a mound of fatty tissue that covers the pubic symphysis   c. is the focus of sexual sensation in the female   d. answers (a),(b) and (c) are correct   e. none of the preceding answers is correct.

26. (LO 17) Each breast consists of   a. about 15 lobes of glandular tissue   b. a system of ducts that open on the surface of each nipple   c. adipose tissue   d. answers (a),(b) and (c) are correct   e. none of the preceding answers is correct

27. (LO 18) Ejection of milk from the breast is stimulated mainly by   a. prolactin   b. estrogen   c. testerone   d. progesterone   e. oxytocin

28. (LO 19) Which of the following techniques are used to diagnose breast cancer   a. mammography   b. breast examination   c. Pap test   d. answers (a), b) and (c) are correct.   e. answers (a) and (d) only.

29. (LO 19) The secretory phase of the menstrual cycle   a. occurs after ovulation   b. is dominated by progesterone   c. includes the days of menstruation   d. answers (a),(b) and (c) are correct   e. answerrs (a) and (b) only.

30. (LO 21) Estrogen   a. stimulates growth of the endometrium   b. is responsible for breast development and other secondary sex characteristics   c. makes cervical mucus thinner and more alkaline   d. answers (a),(b) and (c) are correct   e. answers (a) and (b) only.

31. (LO 22) The stage of life at which the ovaries cease to function is   a. puberty   b. after age 70

c. menopause d. mittelschmerz e. answers (b) and (d) are correct

32. (LO 23) Sexual climax occurs during a. excitement phase b. plateau phase c. orgasmic phase d. resolution e. vasocongestion phase

33. (LO 24) Which of the following are true regarding fertilization a. only one sperm actually penetrates the ovum b. many sperm release enzymes from their acrosomes c. the zona pellucida must be penetrated d. answers (a),(b) and (c) are correct e. answers (a) and (b) only.

34. (LO 20) The sex of the offspring is determined by a. the male alone b. the female alone c. mainly the male, but a certain type of ovum must be present d. equally by both partners e. the type of ovum primarily, but sufficient number of sperm must be present.

(LO 26,27)
35. male method
36. least reliable method listed
37. sterilization procedure
38. mode of action depends upon hormones
39. chemically kills sperm

a. tubal ligation
b. oral contraception
c. intrauterine device
d. condom
e. foam

40. (LO 28) Abortions performed during the fourth month of pregnancy are done by a. saline injections b. suction c. uterine-evacuation (menstrual extraction) d. spontaneous methods e. severing and tying the uterine tubes

41. (LO 29) A primary chancre is the first symptom of a. gonorrhea b. syphilis c. genital herpes d. all of the preceding e. None of the preceding

42. (LO 29) If untreated gonorrhea may lead to a. sterility b. a type of arthritis c. heart damage d. answers (a),(b) and (c) are correct e. none of the preceding answers is correct.

ANSWERS TO POST TEST QUESTIONS

1.d	2.b	3.e	4.a	5.e	6.c	7.c	8.b	9.c
10.e	11.c	12.a	13.d	14.b	15.a	16.d	17.d	
18.c	19.e	20.c	21.d	22.e	23.b	24.a	25.b	
26.d	27.e	28.e	29.e	30.d	31.c	32.c	33.d	
34.a	35.d	36.e	37.a	38.b	39.e	40.a	41.b	
42.d								

19 DEVELOPMENT AND GENETICS

PRETEST

1. Three basic developmental processes are _growth_, _morphogenesis_, and _cellular_ _differentiation_.
2. The outer cells of the blastocyst make up the _trophoblast_ while the cluster of cells which projects into the blastocyst cavity are the _inner cell mass_.
3. On or about the seventh day the embryo begins to _implant_ itself in the wall of the uterus.
4. The embryo is kept moist and protected from mechanical damage by the _amniotic fluid_.
5. The organ of exchange between mother and embryo is the _placenta_.
6. The embryonic tissue layer which gives rise to the nervous system is the _ectoderm_.
7. Muscle and bone develop from the _mesoderm_.
8. The _notochord_ induces the formation of the neural plate.
9. During the seventh month of development the embryo is covered by a downy hair called _lanugo_.
10. The actual birth of the baby occurs during the _second_ stage of labor.
11. During fetal life blood is diverted from the lungs by an opening in the interatrial wall called the _foramen ovale_.
12. Phocomelia and other gross defects have occured in thousands of babies due to the tranquilizing drug _thalidomide_.
13. The chemical substance in the cell which basically stores the genetic information of the cell is known as the _DNA_.

14. Information is transferred from this substance to a messenger substance called __mRNA__.
15. Interacting with ribosomes and tRNA, the instructions incorporated to mRNA produce specific __proteins__
16. DNA consists of phosphate, deoxyribose and these bases: __adenine__, __thymine__, __guanine__, and __cytosine__.
17. Each inactive DNA strand is usually accompanied by a __complementary strand__, and the two of them are wound into a __helical__ shape.
18. The bases of the DNA pair as follows: adenine with __thymine__, guanine with __cytosine__.
19. RNA differs from DNA in that its sugar is __ribose__; and in place of the base thymine it employs __uracil__.
20. If a strand of DNA has base sequence ATTGCATGCTA, its complement must have the sequence __TAACGTACGAT__ and in the case of mRNA, __UAACGUACGAU__.
21. The transfer of information from DNA to mRNA is called __transcription__; from mRNA to protein is called __translation__.
22. A gene is a nucleic acid information sequence that specifies a particular __protein__
23. Many mutations arise because of __substitution__ of incorrect bases in the DNA.
24. It is possible that __differentiation__ of tissues is accomplished by a permanent "locking together" of inappropriate DNA strands by __histone__ protein.
25. Where genes are located __on chromosomes__.
26. a. In the case of a certain pair of flowers, a red and a white, the cross R X W produce offspring (F_1=first filial generation) that are pink. That is because the red parent has contributed a chromosome of the homologous pair governing color, and it bore a gene that was capable of producing __the red__ pigment.
   b. The white parent, on the other hand, LACKED a gene capable of producing any flower color. That was why it was white--it wasn't that it had a gene that made white pigment. Its white color was due to the lack of coloring matter. Thus the white parent contributed a gene (and a chromosome bearing that gene) to the offspring for white color, or more properly, a __defective__ gene for flower color.

16. adenine, thymine, guanine, cytosine (LO 10),
17. complementary strand--helical (LO 10) 18. thymine--cytosine (in that order) (LO 10) 19. ribose--uracil (LO 10) 20. TAACGTACGAT--UAACGUACGAU (in that order) (LO 11) 21. transcription--translation (in that order) (LO 11) 22. protein (LO 11), 23. substitution (LO 13) 24. differentiation--histone (LO 14)

<u>Note</u>: Numbers 25 through 29 are an integrated exercise involving LO 15,16 and 17)

25. on chromosomes 26. a. one--red b. defective, c. one--one--twice--half--half--red--red--pink.
27. one--one--two--unlike--homozygous--heterozygous,
28. red--homozygous--homozygous--heterozygous,
a. dominant--dominant--incompletely dominant,
b. dominant--homozygous, c. experiments--recessive,
29. many--single--alleles,

30. Y(LO 18), 31. X (LO 18), 32. hemoglobin (LO 19), 33. tyrosinase (LO 19), 34. idiocy--triploid--21 (LO 20), 35. XO--underdevelopment (LO 20), 36. X--inactive (LO 21) 37. a. both traits expressed b. alternative form of the same gene on a homologous chromosome c. does not express itself when heterozygous d. unlike members of an allele pair e. the actual expression of genes (LO 22).

PART I: DEVELOPMENT

LO 1: Identify growth, morphogenesis, and cellular differentiation as basic developmental processes, and describe their role in the formation of the organism.

An embryo grows by making new cells by mitosis and by growth of its cells. Through the process of morphogenesis cells arrange themselves to form the shape of the body and its organs. Cells differentiate specializing structurally and biochemically to perform specific tasks.

Using development of specific structures such as the early development of the brain tell how each process identified above is an integral part of its formation.

_____

_____

_____

TEST YOURSELF
Match:

1. cells move about and arrange themselves in a precise manner
2. cells become specialized to perform specific functions

a. morphogenesis
b. mitosis
c. growth
d. cellular differentiation
e. gene transmission

> LO 2: Describe the development of the embryo during its first week of life, and draw and label the stages from zygote through blastocyst.

After fertilization the embryo goes through the cleavage stage of development to become a morula and then a blastocyst. The blastocyst consists of an outer trophoblast and an inner cell mass.

(1) What happens during cleavage?_____

(2) Draw and label each stage from zygote to blastocyst. (Remember that these structures are actually 3 dimensional. It would be helpful to attempt to make models out of clay or play dough.)

TEST YOURSELF
3. The series of cell divisions which partition the zygote into many cells is a. implantation b. cleavage c. ovulation d. placentation e. trophoblast formation

4. An embryo consisting of inner cell mass and trophoblast is a a. morula b. gastrula c. blastocyst d. 16 cell embryo e. cleavage stage embryo

> LO 3: Describe implantation.

On about the seventh day of development, trophoblast cells enzymatically erode away an area of uterine wall and the blastocyst sinks down into the wall, implanting itself there.

(1) Draw a diagram illustrating implantation.

(2) How long does the embryo remain implanted in the uterine wall?_____

TEST YOURSELF

5. Implantation takes place   a. after the placenta has developed   b. in the uterine wall   c. in the wall of the cervix   d. as the sperm enter the ovum   e. when the trophoblast is stimulated by the notochord

> LO 4: Trace the development of the fetal membranes and placenta, and give their functions.

The fetal membranes include the amnion, yolk sac, chorion, and allantois,. The placenta develops from the chorion and the uterine tissue between the chorionic villi.

(1) The two fetal membranes that are considered to be vestigial structures in the human embryo are the _____ and the _____.
(2) Briefly describe and give the functions of
a. amnion_____

b. placenta_____
_____

(3) Describe how materials are exchanged through the placenta.
_____

TEST YOURSELF

6. Which are functions of the placenta: a. brings maternal blood supply and fetal blood supply close together   b. permits exchange of oxygen and nutrients from maternal blood to fetal blood   c. produces hormones   d. answers (a),(b) and (c) are correct   e. answers (a) and (b) only

7. The placenta develops from the   a. amnion   b. yolk sac   c. allantois   d. inner cell mass   e. chorion

The three germ layers (embryonic tissue layers), are the ectoderm, mesoderm, and endoderm.

List the structures which develop from:
   a. Ectoderm_____

   b. Mesoderm_____
   _____

   c. Endoderm_____
   _____

TEST YOURSELF

8. Masoderm gives rise to a. muscle b. lining of gastrointestinal tract c. epidermis d. answers (a), (b) and (c) are correct e. answers (a) and (b) only

9. Endoderm gives rise to a. nervous system b. bone c. blood d. answers (a),(b) and (c) are correct e. none of the preceding answers is correct

> LO 6: Describe the later development of the embryo and fetus.

Almost all organs are laid down during the first month of development. During the second month limb buds develop, giving rise to arms and legs, and testes and ovaries begin to differentiate. During the last 3 months of development the fetus grows rapidly in size.

(1) Identify the time of development at which each of the following occur: (consult Table 19-2 in your text.)
   a. heart begins to beat_____
   b. lanugo appears_____
   c. birth_____
   d. brain begins to develop_____
   e. embryo capable of movement_____
   f. limb buds appear_____
(2) At what time of development then is the embryo usually called a fetus?_____

TEST YOURSELF

10. brain begins to develop
11. heart begins to beat
12. lanugo appears
13. birth

a. 266 days
b. first month
c. last trimester
d. first week
e. 4th month

>LO 7: Identify the events of each stage of labor.

<u>Events of labor</u>: first stage - labor pains, cervical dilation and effacement; second stage - passage of infant through the vagina and birth; third stage - delivery of placenta.

Define:
    a. parturition_____
    b. cervical effacement_____

    c. episiotomy_____
_____

TEST YOURSELF
Match:
14. cervical effacement takes place
15. delivery of placenta
16. delivery of baby
17. longest period of labor

a. first stage
b. second stage
c. third stage
d. fourth stage

>LO 8: Describe the fetal circulation and the changes that take place in the cardiovascular system at birth.

At birth the placental circulation ceases, the umbilical vessels contract, and the ductus venosus and ductus Arteriosus constrict so that blood is rerouted. The foramen ovale is closed by a flap of tissue so that the systemic and pulmonary circulations are completely separated.

Give the function of:

    a. umbilical vessels_____

    b. ductus venosus_____
    c. ductus arteriosus_____
    d. foramen ovale_____

TEST YOURSELF

18. The fetal blood vessel that connects the pulmonary trunk to the descending aorta so that blood is diverted from the lungs is the a. ductus arteriosus b. ductus venosus c. foramen ovale d. umbilical artery e. umbilical vein

> LO 9: Identify common environmental factors that may influence development of the embryo, and give examples of each effect.

Environmental factors that may affect the development of the embryo include nutrition, drugs, oxygen, pathogens, and radiation.

Give the effect that each of the following may have upon the developing embryo:
    a. severe protein deficiency_____
    b. excessive amounts of Vitamin D_____

c. thalidomide_____

d. heroin_____

e. excessive amounts of alcohol_____

f. cigarette smoking_____

g. Rubella_____

h. syphilis_____

i. ionizing radiation_____

TEST YOURSELF

19. Cigarette smoking by a pregnant woman   a. reduces the amount of oxygen available to the fetus   b. may slow growth of the fetus   c. sometimes causes cancer in the fetus   d. answers (a),(b) and (c) are correct   e. answers (a) and (b) only are correct

20. When a pregnant woman is subjected to ionizing radiation   a. there is a higher risk of birth defects b. there is a higher risk of leukemia   c. the infant will probably be born with phocomelia   d. answers (a),(b) and (c) are correct   e. answers (a) and (b) only.

PART II: GENETICS

LO 10: Summarize the function of DNA and of RNA in (a) genetic replication, (b) transcription, and (c) translation.
LO 11: Describe transcription and translation.

DNA exists in a doubly helical molecule of indefinite length. Its backbone consists of alternating deoxyribose units, and from this axis there protrude the nucleotide bases adenine, cytosine, thymine, and guanine. Strand pairs with complementary strand via the bases. A pairs with T, C with G.

In transcription the plus strand of DNA builds an RNA complement to itself. The mRNA then travels to the cytoplasm of the cell, where it encounters ribosomes. Using the tRNA, these last organelles are able to build peptides and proteins.

(1) What three kinds of chemical components comprise DNA?
   a. _____
   b. _____
   c. _____

(2) How are they arranged?
   _____
   _____
   _____

(3) What kind of bases does DNA possess? _____
   a. _____
   b. _____
   c. _____
   d. _____

(4) What are the pairing relationships of the DNA bases?

    a. _____

    b. _____

(5) How does the RNA differ from the DNA? _____

    a. _____

    b. _____

(6) 1. What are the base-pairing properties of DNA?

    a. _____

    b. _____

    2. What are the base-pairing properties of RNA?

    a. _____

    b. _____

(7) Here is the base sequence of a strand of DNA. Give the base sequence of its DNA and RNA complements.

	DNA	RNA		DNA	RNA
A	___	___	T	___	___
G	___	___	T	___	___
G	___	___	C	___	___
C	___	___	A	___	___

(8) Why are ribosomes needed in protein assembly?

_____

_____

_____

(9) How are ribosomes constructed?_____

   a. Physical structure_____

   _____

   b. Chemical structure_____

   _____

(10) What is the relationship between ribosomes and certain antibiotics?_____

   _____

(11) Label each species of nucleic acid in the following diagram and indicate (by number) its place in the information transfer sequence.

   Referring to the letters in the diagram, which correspond to those below

   A. What kinds of nucleic acid occur here?

   <u>DNA</u> sequence number $\frac{1}{2}$
   ___

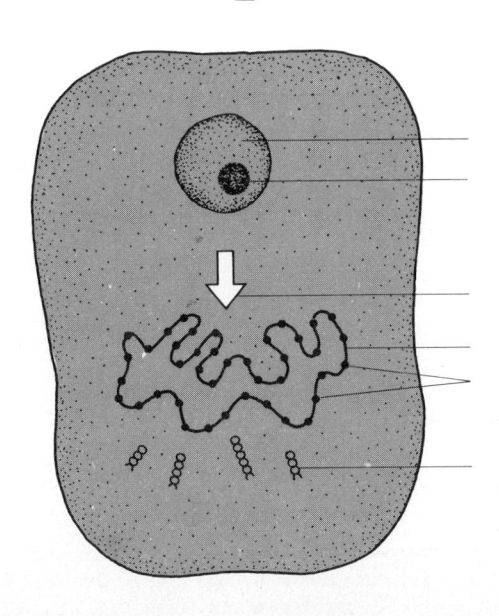

B. Is this structure part of the information transfer sequence?_____ Number_____

C. What substance is this? _____ Number_____

D. What is the name of this membranous structure? _____

E. What are these?_____ What do they do? _____ _____ Number_____

F. What kind of RNA is that?_____ To what is it attached?_____ Number_____

TEST YOURSELF

21. A nucleic acid consists of a chain of nucleotide residues held together by means of   a. peptide bonds   b. phosphate groups   c. glucose   d. pentose molecules   e. carboxyl groups

22. RNA differs from DNA in   a. its phosphate group   b. its pentose group   c. two of its nitrogenous bases   d. that it does not form long strands

23. A piece of mRNA has the base sequence A-U-G. Which of the following would represent the base sequence on the DNA which is complementary:
    a. A-T-C   b. A-T-G   c. U-A-C   d. T-A-C   e. G-C-A

24. Which of the above answers would represent the base sequence on a complementary piece of tRNA?

25. Ribosomes are composed of   a. RNA and proteins   b. three basic particles   c. RNA and lipid   d. tiny outpocketings of the endoplasmic reticulum   e. a membranous envelope surrounding digestive enzymes

26. Ribosomes seem to function, in part, by   a. attaching tRNA to its amino acids   b. breaking the association between mRNA and DNA   c. catalyzing the

formation of the peptide bond   d. producing tRNA, that is, establishing its double stranded nature   e. ensuring that a strand of DNA is "read out" properly

27. Before protein synthesis can take place (choose the best answer).   a. the ribosome must be assembled   b. transfer RNA must be assembled in the nucleus   c. messenger RNA must move down the base sequence of transfer RNA   d. amino acids must attach to tRNA   e. the inducer substance must be removed from the ribosomes

28. Which of the following sequences is correct: a. DNA-mRNA-tRNA-protein   b. DNA-tRNA-mRNA-protein   c. DNA-ribosome-mRNA-protein   d. DNA-mRNA-r-RNA-protein   e. none of the above is correct

LO 12: Characterize and define the gene.

The basic unit of heredity is the gene. Fundamentally, a gene acts by directing the production of a peptide or protein. Characteristics of these proteins determine the traits of the organism.

(1) The basic unit of heredity is the _____, which is basically a set of _____.
(2) A typical gene directs the production of a _____ and specifies its _____ _____ sequence.

TEST YOURSELF

29. Which of the following best defines a gene?   a. a trait of an organism   b. a protein   c. the information needed to produce a protein   d. DNA   e. a kind of body chemistry

LO 13: Define and explain the process of mutation in terms of nucleic acids.

Mutation, i.e. sudden, permanent change in genetic inheritance, may result from some change in the base sequence of a nucleic acid, especially DNA. This may

result in the inactivity, effective absence or other change in the protein product of the affected gene or genes.

(1) Define "mutation" _____

_____

_____

(2) What are some agents that are known to be mutagenic?

_____

_____

(3) What produces most mutations? _____

_____

_____

TEST YOURSELF
---

30. Mutations can occur as a result of the following:
    a. habitual use and disuse of organs   b. certain chemical agents   c. radioactive material   d. x-rays
    e. (b),(c), and (d) but not (a)

31. Most mutations appear to result from   a. certain chemical agents   b. radioactive materials   c. x-rays and cosmic rays   d. body heat   e. no known cause

32. Of the effects given below, the most *fundamental* and basic effect of mutations is   a. producing sterility   b. altering amino acid sequences in protein   c. accelerating the rate of evolution   d. changing the physical characteristics of the organism   e. inactivating enzymes

33. The change of a single new base in a strand of DNA will probably   a. insert an amino acid in the protein product   b. change a substantial number of amino acid residues in the protein product   c. make no difference since inosine is so versatile   c. alter a single amino acid   e. (a),(b) and (d) only

34. Sickle-cell anemia is produced by   a. an inability of the small intestine to absorb iron   b. a single incorrect amino acid in the hemoglobin molecule   c. a recessive gene that affects the bone marrow adversely   d. a pleiotropic effect of the genes involved in determining skin color   e. antibodies that act against one's own red blood cells

> LO 14: Summarize the authors' proposed explanation of the differentiation of cells and tissues.

Though body cells are presumed to share the same heredity, they differ greatly from from one another in shape and in function. That implies that a means of supressing the expression of inappropriate genes must exist. The suppression may be accomplished by a permanent or temporary locking together of the DNA strands in the area of the inappropriate genes, and perhaps the nuclear histones(and acidic receptor proteins) are involved in the process.

(1) The ancestry of all body cells may be traced to a single _____ from which they have descended by mitosis.
(2) The process of mitosis produces cells that are genetically _____.
(3) The zygote must possess all the genetic information utilized by all body cells. Specialized cells must therefore _____ some or most of this information.
(4) In addition to DNA, cell nuclei possess alkaline proteins called _____. These may be able to prevent DNA from uncoiling. This would prevent _____ from taking place

TEST YOURSELF

35. Which of the following has been suggested as repressing inappropriate genetic information in different body cells?   a. mRNA   b. histone protein   c. tRNA   d. rRNA   e. serotonin

LO 15: Relate the inheritance of genetic traits to the behavior of chromosomes in meiosis.

Genes are associated with chromosomes and occur on them. The behavior of genes is similar to the behavior of chromosomes. Adult persons possess both chromosomes of a given pair, i.e. they are diploid. Gametes, however, are haploid, i.e. each gamete possesses but one member of each chromosomal pair.

Normally no offspring can receive more or less than one representative of a homologous pair of chromosomes from each parent. The member of a pair of chromosomes or of their contained genes that a gamete receives is governed entirely by chance. The particular combination of maternal and paternal chromosomes received by an offspring is therefore also governed by chance.

The inheritance of a particular chromosome has no influence on the inheritance of any other chromosome not homologous to it. The inheritance of any gene has no influence on the inheritance of any other gene borne on a nonhomologous chromosome.

(1) Complete the following exercise:
Genes which do not govern the same trait are transmitted to offspring independently of one another. Today we know that this will be true only if they are located on *different* chromosome pairs.

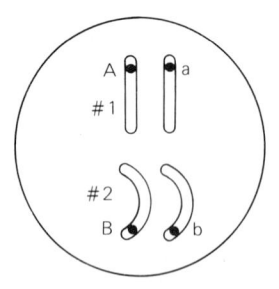

Since offspring are derived from zygotes and zygotes originate from the union of gametes, genes must be transmitted from parents to offspring via gametes (sperm, and eggs). Meiosis separates chromosomes so that each gamete receives *one* representative of *each* pair. But which one?

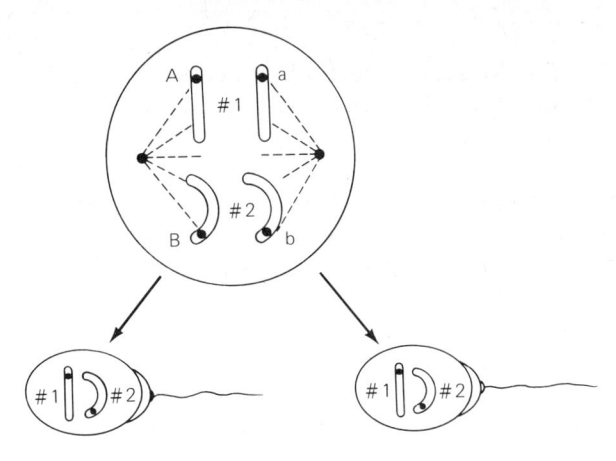

Each gamete *must* receive a No. 1 chromosome and a No.2 chromosome, and indeed one representative of each pair of homologous chromosomes. A sperm may receive the chromosome bearing A or the chromosome bearing a, and it must receive one of them - but *not* both of them. Similarly, it may receive either the chromosome bearing B or the chromosome bearing b, but *not* both of them. However, it must have one of the No. 1 chromosomes and one of the No. 2 chromosomes.

Each sperm has an *equal* chance of getting either member of a homologous chromosome pair. Expressed as a percentage, what is the chance that a sperm will receive the chromosome bearing gene A?_____
a?_____

B?_____ b?_____

The particular member of the No. 1 chromosome pair that a given gamete receives has no bearing upon which representative of the No. 2 chromosome pair it will receive.

TEST YOURSELF

36. Suppose an organism has 3 pairs of chromosomes, I, II, and III. Each pair consists of a paternal member and a maternal member. The organism now reaches sexual maturity and produces gametes. Which of the following chromosonal combinations will normally NOT occur in one of those gametes? a. MI, PII, MIII b. MI, PI, MIII c. MI, PII, PIII d. MI, PI, PIII e. all of the above are easily possible

LO 16: Summarize the concepts of homologous
chromosomes and allelic genes.

The production of any protein or peptide is governed
by two genes present in each and all body cells. These
genes are alleles. Allelic genes are carried on
homologous chromosomes which, like them, are paired.
One member of each homologous pair of chromosomes is
maternal in origin. The other member is paternal in
origin.

(1) Study the illustration on the following page.

(2) What determines whether genes are allelic?

_____

(3) How many genes of an allelic kind can a human cell
possess?

_____

(4) Are allelic genes always the same?

_____

TEST YOURSELF

37. Which of the following chromosomes are homologous?
a. MI, MII b. MI,PII, c. MI PI   d. MI, MII,MIII

38. Which of the following genes are probably allelic?
a. normal color vision and color blind b. nearsightedness and color blind c. congenital cateract
and color blind d. all of the preceding e. (b)
and (c) are both correct choices

LO 17: Distinguish between homozygous and
heterozygous genotypes. Relate genotype to
phenotype in terms of dominance.

Members of a gene pair may be alike (homozygous) or
unlike (heterozygous). When members of a gene pair are
unlike, the traits which they govern may in some cases
be determined by one of them if it is clearly dominant,
(continued on p. 394)

Every eukaryotic cell has chromosomes, normally visible only during cell division. These chromosomes consist in part of DNA and bear upon them, in the form of configurations of that DNA, the genes that govern all or most inherited traits. The members of a given chromosome pair are said to be *homologous*

GENE _____

Each chromosome presumably is made up of perhaps thousands of genes. The few genes that we have discovered occupy definite physical locations on the chromosomes known as *gene loci*

These genes are _____

These genes are *not* allelic to one another

Since diploid organisms possess pairs of homologous chromosomes, the genes borne in corresponding loci of the pair also occur in pairs. If genes occupy the same locus on each of a pair of chromosomes, they are said to be *allelic*. Allelic genes always govern the *same trait*

Gene for seed color (purple)

Gene for seed color (yellow)

However, even though allelic genes govern the same traits, they need not necessarily contain the same

_____

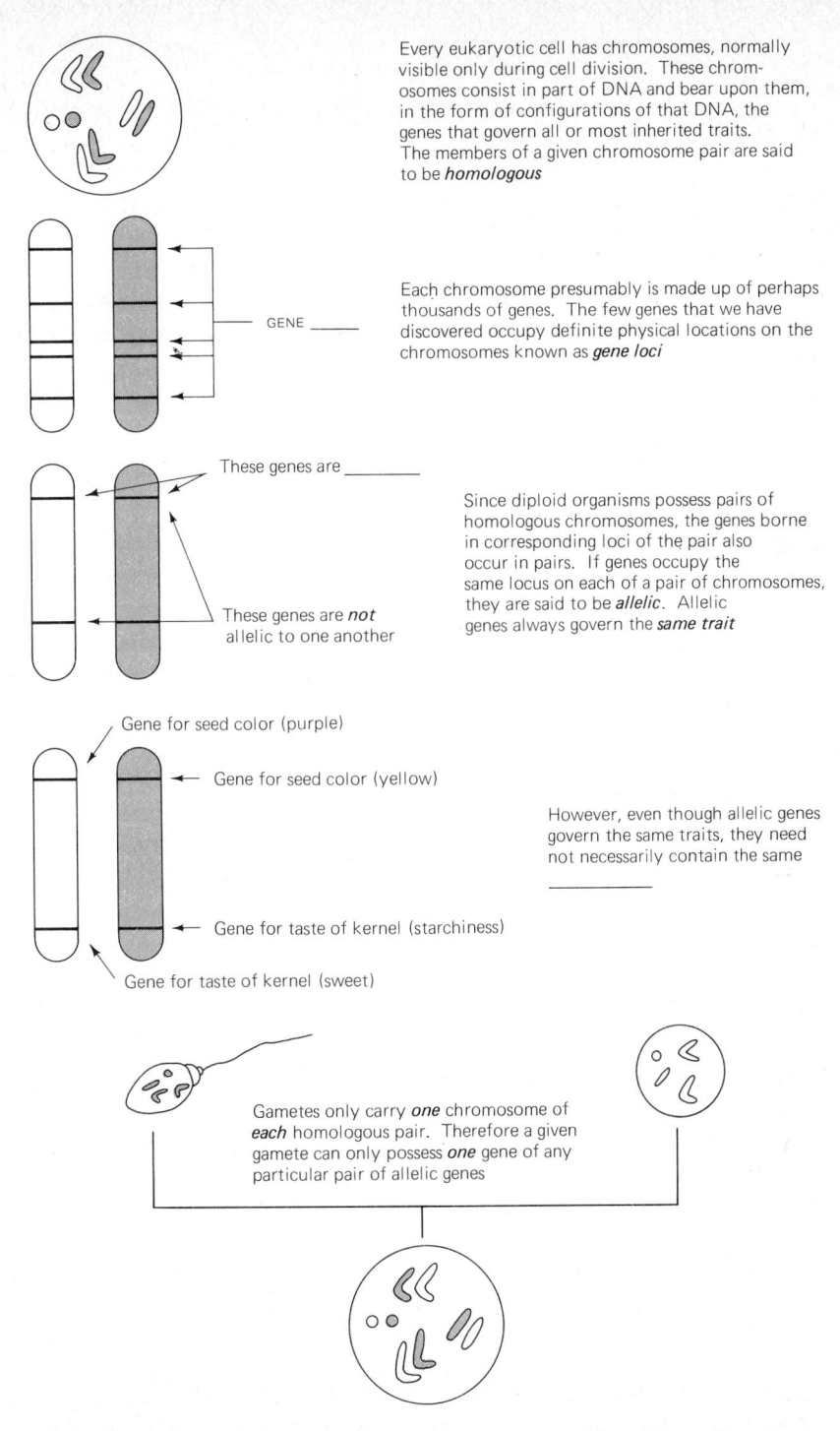

Gene for taste of kernel (starchiness)

Gene for taste of kernel (sweet)

Gametes only carry *one* chromosome of *each* homologous pair. Therefore a given gamete can only possess *one* gene of any particular pair of allelic genes

When the gametes combine to form a zygote, it, and the resulting embryo, will have homologous pairs of chromosomes, but of each pair, one member will be MATERNAL in origin, and one PATERNAL in origin. Each pair will bear allelic genes.

and in others, where dominance is not clear-cut, by both of them. A gene that is expressed only when homozygous is said to be recessive.

(1) Review the following important facts:
   a. The hereditary factors are called *genes*. The genes possessed by an organism comprise its *genotype*.
   b. The expression of heredity depends upon the genes that are present but is not necessarily the *same* as those genes.
   c. The physical or other expression of the genes is called the *phenotype*.
   d. There is *no* central storehouse of genetic information in the organism. *Each cell* possesses *all* the genes of the entire body. In particular, this information is found in the *nucleus* of each cell. The nucleus of the cell contains *chromatin*, which during cell division is packaged into *chromosomes*. Each cell therefore possesses all the genes of the organism, and these genes occur in the chromosomes.
   e. Sex cells carry *one* chromosome of *each* pair. Upon fertilization, the pairs form again, one member of each pair coming from the sperm (paternal chromosome) and one member from the egg (maternal chromosome).

(2) Answer the following questions: (Several of which are review questions).
   a. What is meant by dominant?_____
   Recessive?_____
   Condominant?_____
   b. What is a genotype?_____
   c. What is a phenotype?_____
   d. What are homologous chromosomes?_____
   _____

e. What are alleles?_____

_____

f. What is meant by homozygous?_____

_____

heterozygous?_____

_____

TEST YOURSELF

39. A white guinea pig is mated to a black. All offspring are black. The genetic principle that applies to this example is   a. the law of segregation   b. the principle of linkage   c. the principle of dominance   d. the law of independent assortment

40. In the example of No. 39, suppose one of the offspring had been white   a. the white offspring is homozygous   b. the black parent is heterozygous   c. the white parent is homozygous   d. all of the preceding are true   e. the black parent could have been heterozygous or homozygous

41. The white guinea pig born to those two parents   a. got both white genes from the white parent   b. got one white gene from the black parent   c. may have gotten both white genes from the black parent   d. all of the preceding are possible

> LO 18: Describe the inheritance of sex-linked genes.

The X chromosome is exceptional in that it has no homologous mate in the male. A gene located on the X chromosome will be expressed if it is dominant in the female, but regardless of dominance in the male.

Answer the following questions.
(1) What determines sex?_____
_____
_____

(2) How does the Y chromosome produce its effect?
_____
_____
_____

(3) Why is X chromosome able to carry many genes that seem to have nothing to do with sex, while the Y chromosome, apparently, is not able to?
_____
_____

(4) What is a sex-linked gene? _____

With which chromosome is it usually associated? _____

List three sex-linked traits: _____
_____

(5) How are most red-green color blindness and hemophilla inherited? _____
_____
_____
_____

TEST YOURSELF

42. A male person has   a. no X chromosome   b. two Y chromosomes   c. two X chromosomes   d. no Y chromosome   e. an X and a Y chromosome

43. Few genes are Y-linked. The reason for this is probably that   a. the Y chromosome is largely homologous with the X   b. both sexes possess two Y chromosomes   c. both sexes possess two X chromosomes   d. the Y chromosome occurs in only one sex

LO 19: Summarize the characteristics of selected genetic diseases.
LO 20: Summarize the characteristics of selected chromosomal disorders.

Please consult table 19-4 and 19-5 in the textbook
_____
_____

TEST YOURSELF

Symptoms	Possible cause
44. cat like cry, idiocy	a. psuedohypertrophic muscle distrophy
45. sexually underdeveloped	b. abnormal hemoglobin
46. abnormally small red cells	c. Turner's syndrome, XO
	d. incompetent plasma globulin
47. inability of blood to clot	e. shortened chromosome No. 5

LO 21: Summarize the Lyon hypothesis of the inactivation of X chromosomes.

The Lyon hypothesis proposes that, of every pair of X-chromosomes in each cell of the human female, one is randomly inactivated.

(1) What is a Barr body? _____

(2) Summarize the Lyon hypothesis _____

(3) What is Turner's syndrome? _____

(4) What karyotype is responsible for it? _____

(5) What is the effect of XYY karyotype? _____

TEST YOURSELF

49. Down's syndrome (mongoloid idiocy) is usually the result of   a. an XXY genetic constitution   b. an XYY genetic constitution   c. an XO genetic constitution   d. a triploid No. 21 autosomal chromosome

50. An apparently male infant with a catlike voice and severe mental retardation, proves (upon microscopic examination of his blood) to have   a. an XYY genetic makeup   b. a shortened No. 5 chromosome   c. a triploid No. 23 chromosome   d. a completely haploid chromosome makeup

51. The Y chromosome   a. is essential to life   b. is absent in persons with Down's syndrome   c. is inactivated in 50 percent of the body cells   d. determines sex

> LO 22: Be able to define the basic terms relating to genetic inheritance--for example, gene, dominance, recessiveness, codominance, chromosome, homozygous, hetrozygous, alleles, homologous, genotype, and phenotype.

We suggest that you obtain   3 x 5-inch file cards, write the term on one side of each card, and the definition on the other. By turning the card over you can find out whether you know the definition of each term.

ANSWERS TO ALL TEST YOURSELF QUESTIONS

1.a	2.d	3.b	4.c	5.b	6.d	7.e	8.a	9.e
10.b	11.b	12.c	13.a	14.a	15.c	16.b	17.a	
18.a	19.e	20.e	21.b	22.b	23.d	24.c	25.a	
26.a	27.d	28.a	29.c	30.e	31.e	32.b	33.d	
34.b	35.b	36.b	37.c	38.a	39.c	40.d	41.b	
42.d	43.d	44.e	45.c	46.b	47.d	48.a	49.d	
50.b	51.d							

POST TEST

1. (LO 1) Which of the following is an example of morphogenesis: a. multiplication of cells during cleavage   b. growth of individual cells during blastocyst stage   c. movement of cells of the neural plate to form neural folds   d. specialization of heart cells to contract   e. differentiation of neural cells in response to the notochord

2. (LO 2) During cleavage: a. the tiny heart of the embryo begins to beat b. limb buds form c. the placenta becomes functional d. the zygote is partitioned into many small cells e. the embryo implants itself in the uterine wall

3. (LO 2) The morula consists of a. trophoblast b. inner cell mass c. a tiny cluster of about 16 cells d. answers (a),(b) and (c) are correct e. answers (a) and (b) only are correct

4. (LO 3) The stage at which the embryo implants in the uterine wall is a. cleavage b. blastocyst c. morula d. zygote e. neural tube stage

5. (LO 4) Keeping the embryo moist is a function of the a. amnion b. yolk sac c. allantois d. chorionic villi e. placenta

6. (LO 4) Which are functions of the placenta a. nutrition of the embryo b. providing oxygen for the embryo c. removal of wastes from the embryo d. answers (a),(b) and (c) are correct e. answers (a) and (b) only.

7. (LO 5) The ectoderm gives rise to a. brain b. epidermis c. bone d. answers (a),(b) and (c) are correct e. answers (a) and (b) only

8. (LO 6) Which of the following events occur during the first month of development a. heart begins to beat b. nervous system begins to develop c. gastrointestinal and respiratory systems begin to develop d. answers (a),(b) and (c) are correct e. none of the preceding answers is correct

9. (LO 7) During the first stage of labor, a. cervix is dilated b. cervical effacement takes place c. the placenta is delivered d. answers (a), (b) and (c) are correct e. answers (a) and (b) only

10. (LO 8) Which of the following are structures (vessels) found in the embryo but not in the normal adult: a. foramen ovale b. ductus venosus c. ductus arteriosus d. answers (a),(b) and (c) are correct e. answers (a) and (c) only

11. (LO 9) Which of the following environmental influences have been shown to cause birth defects: a. excessive use of vitamin D b. ionizing radiation c. severe protein malnutrition d. answers (a), (b) and (c) are correct e. none of the preceding answers is correct

12. (LO 10) The fundamental genetic material of the cell, in which all genetic instructions are registered, is: a. MRNA b. tRNA c. histone protein d. DNA

13. (LO 11) Transcription involves a. construction of mRNA complementary to DNA b. assemblage of tRNA units c. "readout" of mRNA by the ribosome d. production of protein e. (c) and (d) only

14. (LO 12) Which of these could be a gene? a. a specific amino acid sequence b. the biochemical product of a given enzyme c. a sequence of bases in a DNA strand d. a kind of mRNA e. a sequence of tRNA species

15. (LO 13) A mutation may occur a. as a result of ionizing radiation b. when an incorrect base occurs on a DNA strand c. due to the action of certain carcinogenic chemicals d. undetected in a heterozygous carrier e. all of the preceding are correct

16. (LO 14) Histone proteins may a. actuate tRNA b. bind DNA strands together c. damage mRNA d. produce polyribosomes e. be a part of the cell membrane

17. (LO 15,16) Allelic genes, a. occur on homologous chromosomes b. occupy corresponding loci c. govern the same traits d. are never more than two in number e. all of the preceding are true

18. (LO 17) An albino man marries an apparently normal woman. The children appear to be normal. Which of the following is certain to be true? a. the children are homozygous b. the children are heterozygous c. the man is homozygous d. the woman is homozygous e. (b),(c) and (d) are true

19. (LO 18) A gene located on the X chromosome a. will always be expressed even if it is recessive b. can be expressed only in males c. will only be expressed if it is dominant d. can be expressed even when it is alone

(LO 20-23)
20. X-linked recessive trait
21. Leads to COPD and digestive malfunction
22. Abnormally tall stature
23. Testes slowly degenerate

a. XYY
b. Down's syndrome
c. Klinefelter's syndrome
d. cystic fibrosis
e. hemophilia

24. (LO 20) Jack is exceptionally tall and has an IQ of 90. Which of the following karyotypes might he have: a. shortened No. 5 b. XYY c. XO d. XXY e. trisomy 21?

25. (LO 21) The "drumstick" chromosome often found in the nuclei of white blood cells a. indicates the sex of the person involved b. represents an inactivated Y-chromosome c. could occur in a person with the XYY syndrome d. is characteristic of the cri-du-chat disorder

ANSWERS TO POST TEST QUESTIONS

1.c   2.d   3.c   4.b   5.a   6.d   7.e   8.d   9.e   10.d
11.d 12.d 13.a 14.c 15.e 16.b 17.e 18.e
19.d 20.e 21.d 22.a 23.c 24.b 25.a

APPENDIX TO CHAPTER 19

HOW TO SOLVE GENETIC PROBLEMS

Your instructor may wish to test your mastery of genetic principles by assessing your ability to solve genetic problems. You may, in the future, have occasion to council others regarding genetic diseases they may have which they may possibly carry in a heterozygous state. To do this intelligently, you must have the ability to solve practical genetic problems.

In this appendix, we provide samples of approaches to the most common kind of practical genetic problems. In each case study the principle involved, then go through the worked example carefully, making sure that you understand it. Then do the unworked examples. Hints and answers are at the end of this appendix

Additionally, you should try to answer the study questions provided, for they will help you to assess your understanding of the principles involved.

1. Read the problem.
2. Determine what traits are dominant and which are recessive. Often you must marshall background knowledge to do this -- which may not be explicitly mentioned in the problem.

3. Are any letters assigned to the genes? If not, make some up. We usually take the *dominant characteristic* and use the first letter of that word. For example, if *polydactyly* (extra fingers) is dominant over the normal five fingered condition, we would pick P for the dominant gene, and small p for its recessive normal allele. You may, if you wish, use different letters (as P for polydactyly and N for normal) for different allelic genes, but the practice is sometimes confusing.

4. Determine, if possible, the genotypes of the parents. In 9 out of 10 problems this information is given, or at least implied. Sometimes you have to

deduce it from other information given. Write it down so that you can remember what it is.

5. Determine *all the possible* kinds of gametes that can be made by each parent. Be careful, remember that a gamete can ordinarily receive only one gene of a pair of alleles. This is the part that most people have trouble with!

6. Make a Punnett square, using each of the gametes for one parent across the top of a column and the head of the horizontal row. If you have done step 5 properly you shouldn't have any trouble with this step.

7. Work the cross carefully.

8. Now read the problem again. Find out exactly what it is asking for. Don't assume too much. This is another place where many people get lost.

9. In most problems, these steps should get you through adequately. Some are slightly altered--for example, if the genotype of one of the parents is unknown, and that is what the problem wants you to discover. You might assign that parent something like A? or ?? genotype and see if that helps. Put the offspring genotypes in the square and work backward. Remember, this won't get all the problems--there is nothing like real understanding--but it can help organize your attack on a genetics problem. And of course, unless you understand the terms, such as homozygous, heterozygous, dominant, recessive, allele, and so on, you can't begin to think of working problems.

10. Finally, the actual genetic information you need to solve these problems often appears concealed rather than revealed by the wording of the problem. Learn to translate such a sentence "Mary is normally pigmented but had an albino father" into its logical consequence: "Mary is heterozygous for albinism" and then into "Mary is Cc". Notice that, in this kind of a problem you may need to solve several subsidary problems before you can proceed with the final solution. Work them systematically one at a time.

I. SIMPLE MONOHYBRID CROSSES

LO 22: Given parental genotypes and dominances of the genes involved, solve simple monohybrid crosses involving a single pair of alleles.
   a. When the parents are heterozygous

PRINCIPLE: In a monohybrid cross, one investigates the behavior of a single pair of allelic genes by determining (1) the genetic makeup of all possible gametes produced by both parents (proceeding on the assumption that only one member of an allelic pair of genes may be distributed to each gamete), and (2) all possible combinations of the genes of the maternal gametes with the genes of the paternal gametes, and (3) the phenotypes and proportions that are predictable among the offspring on this basis.

Required information: (1) dominance and recessiveness of the genes involved (2) genotypes of parents.

EXAMPLE 1: Jack and Jill are both heterozygous for albinism, a recessive trait. What proportion of their offspring will be albino?

Given: Albinism is recessive, therefore normal coloration must be dominant. Both parents are heterozygous.

Let A stand for the normal allele, and "a" for albino gene. If the parents are heterozygous, that is genetically impure for this trait, then each of them must have both gene A and gene a. One can have no more than two genes for each allele, so their genotypes are clearly Aa, in both the cases.

Now we must find the kinds of gametes each can produce. Since only one member of an allelic pair of genes can enter a gamete:

Jill (Aa) produces gametes with A and other gametes with a. Jack (Aa) produces gametes with A and other gametes with a.

We can work out all possible combinations like this:

```
            ♂ gametes
               A        a
               A        a

                ┌───────┬───────┐
                │  AA   │  Aa   │
             A  │       │       │
                │       │       │
♀ Gametes       ├───────┼───────┤           You can fill
                │       │       │    ←──    these in
                │       │       │
             a  │       │       │
                └───────┴───────┘
```

Since A is dominant, all cases of AA and of Aa will be phenotypically normal. Only aa will be actually albino. This will make up about 1 out of 4 of the offspring, so the answer is 25%.

MODEL:

Information given: _____
_____
_____

Genotypes of parents: (Male) _____ and _____
                     (Female) _____ and _____

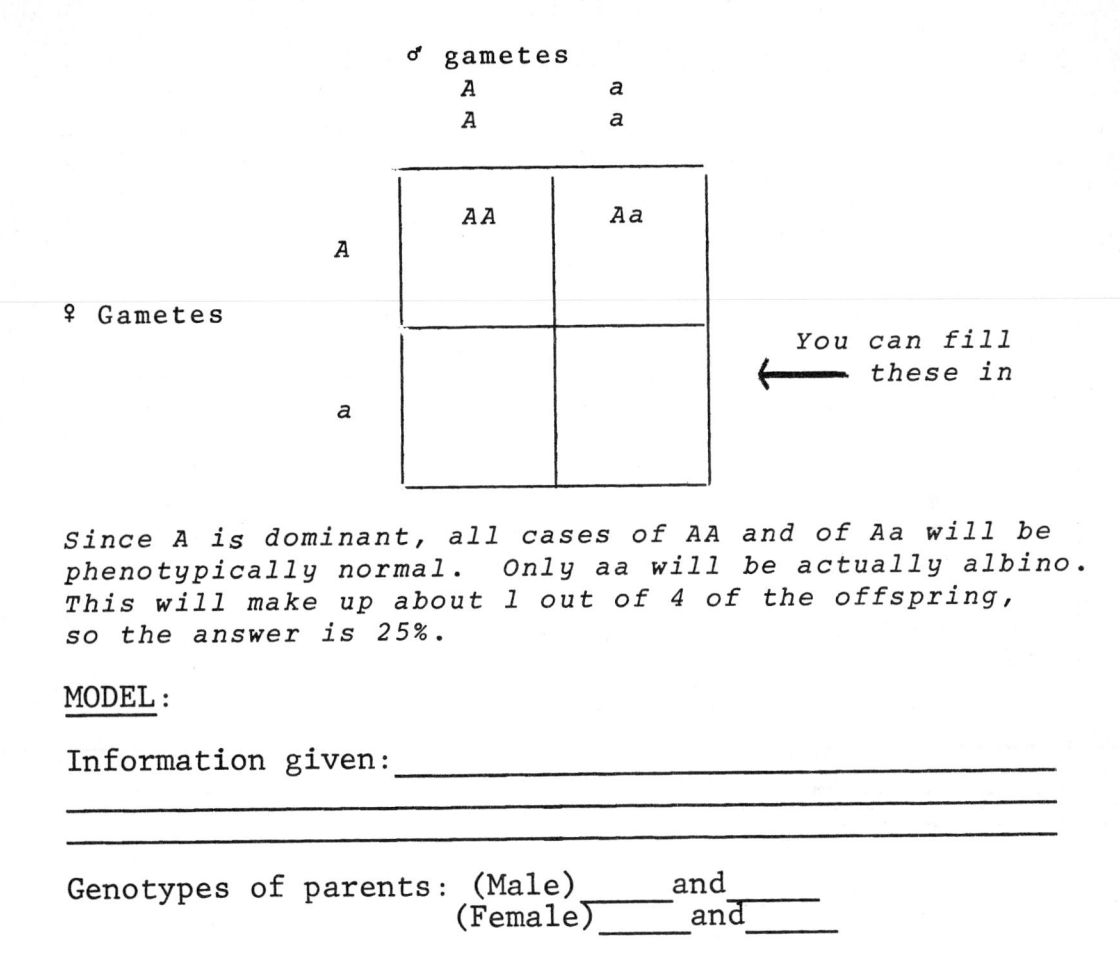

Possible combination of gametes:

                                          ♂ gametes

            ♀ gametes

Phenotypes: _____ , _____ , _____
Conclusion _____

EXAMPLES FOR YOU TO DO (See p. 423 for help)

(1) A gardner buys red and white lupines. To save money, he plants the seeds they produce. Some of them produce pink flowered plants. Wanting more pink plants, the gardner crosses several of these and plants the seeds. When they mature, he discovers to his amazement that one fourth are red, one fourth white, and one half pink. Assuming that red and white are imcompletely dominant in lupines, show why this has occurred.

(2) Sickle cell anemia is a recessive, very serious genetic illness. Ruby and Paul are heterozygous for this trait. Will any of their children be phenotypically normal? What proportion?

(3) Wilson's disease results from the abnormal accumulation of copper in the body, leading to liver or kidney damage. The drug penicillamine must be given to help the body excrete the copper. Jeff and Josie are heterozygous for this trait. What are their chances of having a child who exhibits it? Assume that Wilson's disease is recessive.

(4) White forelock is a dominant trait. Two first cousins with white forelock marry. Assuming they are both heterozygous for it, what is the chance that their child will eventually develop a white forelock?

> LO 22b: When one parent is homozygous and the other is heterozygous.

EXAMPLE 2 A breeder crosses a heterozygous black guinea pig with a white one. If black is dominant, predict the phenotypes and proportion of the offspring.

Given: One parent heterozygous black.
Other parent white. Black is dominant.

Let B stand for black, b. for white. Really it is obvious that black is dominant - for if it were not, it would not show up at all in a guinea pig that is heterozygous for it. Clearly the black pig, then, is Bb. The best definition for a recessive trait is that it is expressed only when it is <u>homozygous</u>. Therefore, the white pig must be bb.

Since only one member of an allelic pair can enter a gamete, the black parent (Bb) can produce TWO kinds of gametes, B and b. But the white parent (bb) can only produce ONE kind of gamete. Since both its color genes are the same, this gamete type will be b only.

To work out the possible combinations we need only a two-cell Punnett square:

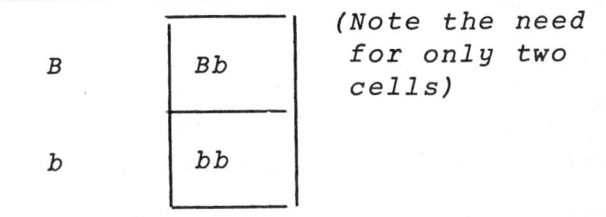

	B
B	Bb
b	bb

(Note the need for only two cells)

50% of the offspring will be white, 50% black.

EXAMPLE 3: A form of gout is inherited as a recessive trait. In this disease, excess uric acid builds up in the blood, and is deposited as tiny crystals in the soft tissues of the limbs and around the joints. This causes very painful swelling somewhat resembling acute arthritis. The disease can be treated with the drug allopurinal. Joe and Belinda marry. Joe is genetically normal, and Belinda is heterozygous for the disease. How likely are their children to have hereditary gout?

Given: Gout recessive. One parent only is heterozygous for the trait.
Joe is GG. Belinda is Gg. Joe can produce only G gametes. Belinda can produce two kinds of gametes, however, G and g. Thus:

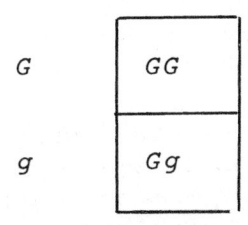

	G
G	GG
g	Gg

Conclusion: All offspring phenotypically normal.

MODEL:

Information given:_____
_____

Genotypes of parents:_____and_____

Possible gametes_____,_____,_____

                                                                        (Note the need for only two cells)

Phenotypes:_____,_____

Conclusion:_____

EXAMPLES FOR YOU TO DO:

(5) If red and white are codominant in four o'clocks, what will be the result if we cross a pink flowered plant with a red? With a white?

(6) Intestinal lactase deficiency involves the absence of an enzyme that will digest lactose, the chief sugar found in milk. As you might expect it is a recessive trait, and can be controlled simply by withholding milk products from the diet. Failure to do so will produce severe diarrhea. Janet Wong and Kim Lee marry. Kim has the lactase deficiency and Janet is heterozygous for the trait. What is the chance that their first child will have the disease?

      LO 22c: When some or all of the parental
      genotypes must be inferred.

EXAMPLE 4: One form of rickets is inherited as a recessive trait. It can be treated by large doses of vitamin D. Geoffrey has this form of rickets, which drew him and Ermintrude together. Erm's father had it, also, you see though her mother was perfectly normal as were all ancestors on her mother's side. What is the likelihood that their children will escape the disease?

Given: Trait is recessive, Geoffrey has disease.
      Ermintrude does not, but her father did.

Geoffrey is obviously rr. Erm could be either RR or Rr judging from her normal phenotype alone, but since her father had the disease, and since one must receive one member of each allelic pair of genes from one's mother, and one from one's father, she must have one r gene from her Dad. Geoffrey, of course, was homozygous for the trait--that's the only kind of ricket-gene that he had. The cross, therefore, works out to Rr X rr. The rest, hopefully, is obvious.

MODEL:

Since this next type of problem is similar in principle to examples 1, 2 and 3, there is no need to give a separate model for it. However, there is one additional step, inferring the genotype of one or both parents from information about <u>their</u> parentage. Thus, in this type of problem one must perform two or three crosses rather than one:

(1) A cross to determine the parental genotype from that of the grandparents. In example four this was RR x rr ⟶ Rr.
(2) The parental cross. In example four this was Rr x rr.

EXAMPLES FOR YOU TO DO:

(7) Methylmalonic acidemia is a recessive trait that produces idiocy when homozygous. It can be treated with massive doses of certain B-vitamins. Chris and Russ are married. Their second child has the disease. Russ dies and Chris remarries a completely normal man whose ancestry appears genetically normal. What is the chance that <u>their</u> child will have the disease?

STUDY QUESTIONS

1. What is a pair of allelic genes?
2. What is the reason that only one number of each pair of alleles can be distributed to a gamete?
3. What is meant by dominant? By recessive?
4. Why can't you have more than two members of an allelic pair of genes in the body cells?

5. Do I have a right to assume that a gamete from the worked example, could only be A or a? Why not AA, aa or Aa? (p. 405)
6. Must a male gamete always combine with a female gamete? Why not a male with a male?
7. What is meant by phenotype? If I know which gene is dominant, can I be sure of the phenotype of an Aa individual? Can I be sure of the genotype of an albino? Of a normally colored person?
8. What is codominance? Incomplete dominance?
9. In the case of a recessive genetic disease which of these will exhibit the trait: homozygous normal, heterozygous or homozygous recessive?
10. Which of these will exhibit a dominant trait: homozygous dominant, heterozygous, or homozygous recessive?
11. Why should it be obvious that the white pig is homozygous? (p. 406)
12. Why should we use a two-cell square in this problem rather than one with four cells? (p. 407)
13. Let's suppose that red were dominant (as it is in pea plants). What phenotypes and proportions would example 5 produce then? What determines the degree of dominance in a case like this?
14. Why did I give the people of example 6 Chinese last names? Can you find out?
15. Ordinary rickets can also be treated with vitamin D, and it is not hereditary. What, then, is the difference between the hereditary and acquired disease?
16. How many crosses did you have to perform to get the answer to example 7?
17. Is it reasonable to assume that the apparently normal people in example 7 were, indeed, genetically normal?
18. If you were a genetic counselor, would you advise Chris' children not to have children of their own? (p. 409)

II. MULTIPLE ALLELES AND BLOOD TYPES

LO 23: Describe the inheritance of the ABO and Rh factors, outline the mechanism of Rh disease, and solve problems involving the inheritance of these traits.

INTRODUCTION: Inasmuch as the textbook mentions the ABO and Rh series of blood types in chapter 14, we suggest you review that section before attempting to work blood type problems.

The genetics of the ABO series is unusual. Though one may have only <u>two</u> alleles governing these blood types, <u>three</u> exist, if you count type O. O, being a <u>lack</u> of antigen is recessive to both A and B, so that type A blood can represent the gene type AA or AO, and type B blood similarly is either BB or BO. But only the genotype AB can produce AB blood, and of course, only OO can produce type O blood.

We will assume for our present purposes that there are only two Rh blood types, positive and negative, (though this is an oversimplification). The Rh negative phenotype lacks antigens of the Rh series. Rh-positive (R) is dominant over Rh-negative (r).

1. Fill in the blanks:

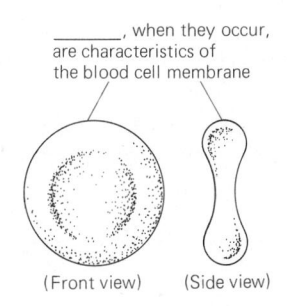

_____, when they occur, are characteristics of the blood cell membrane

(Front view)   (Side view)

Red blood cells

_____, when they occur, naturally are present in the blood plasma

411

1. a With respect to the ABO series of blood types, what are antigens? _____
_____

   b. Where do they occur? _____

   c. What are antibodies? _____

   d. Where do they occur? _____

   e. Can you deduce a general principle governing the occurence of an antibody and antigen of the ABO series? _____
_____

2. a Complete this table:

Phenotype	Genotype(s)	Antigens	Antibodies
A	_____	_____	_____
B	_____	_____	_____
AB	_____	_____	_____
O	_____	_____	_____

   b. Summarize the inheritance of ABO blood type. _____
_____
_____

3. In blood of what type are Rh antigens naturally associated with the cell membranes? _____
4. In blood of what type (if any) are Rh antibodies naturally associated with the plasma? _____
5. Who may acquire anti-Rh antibodies? _____

   How? (a) _____

   (b) _____
_____

412

6. How is Rh blood type inherited? _____
   _____

7. Describe the cause of erythroblastosis fetalis.
   _____
   _____
   _____
   _____

PRINCIPLE: Crosses involving Rh factor are handled, in principle, much as is any other monohybrid cross and therefore require no special discussion. Indeed this could be said also for crosses involving the multiple alleles of the ABO system. The possibility of confusion, however, is greater and must be carefully guarded against. One (1) must determine the possible genotypes of all persons involved and (2) usually must work the problem several ways unless (3) we can eliminate some of the possibilities by logic.

EXAMPLE A woman with type O blood claims a certain man is the father of her child. The child also has type O blood. The man is found to have type AB. Can he be the father of this child? Support your answer with an explanation of the genotypes involved.

Given: Woman is type O, child is type O, accused father is type AB.

Background knowledge required: O is recessive to A and B. A and B are codominant.

Find: possible genotypes resulting from the cross O X AB

The woman, being type O could only have the genotype OO. The man could only have the genotype AB. Such a cross could therefore be shown as:

	A	B
O	AO	BO

The resulting phenotypes will be A and B, since O is recessive.

It is also possible to solve this problem by a more abstract logical approach. A type O baby has an OO genotype. Since one member of a pair of alleles always comes from one parent and the other from the other parent, the father as well as the mother would need to possess at least one O gene in his genotype. Thus he could be OO, BO, or AO but <u>not</u> AB.

MODEL:

Information given_____

_____

_____

Genotypes of supposed parents:_____and_____

Possible gametes of supposed parents:
    (male)_____and_____
    (female)_____and_____

                                    or

Possible genotypes of offspring_____ _____ _____
Possible plenotypes of offspring_____ _____ _____

Does the child have one of these phenotypes or genotypes?_____yes_____no

Conclusion_____

_____

_____

EXAMPLES FOR YOU TO DO   (See p. 424 for help)

1. Frankie and Johnny were lovers. Frankie's blood type was AB and Johnny's was A. Frankie had a baby, and according to the hospital its blood type was O. What are the implications?

2. Mrs. Wong and Mrs. White had their babies on the same day; the hospital mixed the offspring up and could not tell them apart. The doctors had blood tests performed on all concerned so as to straighten the situation out.

Here are the findings: Mrs. Wong: A, Rh^+
Mr. Wong: AB, Rh^+
Mrs. White: AB, Rh^-
Mr. White: AB, Rh^-
Baby girl: A, Rh^+
Baby boy: B, Rh^-

a. Girl baby is Wong, boy baby is White. b. Boy baby is Wong, girl baby is White. c. Either couple could have produced either baby. d. neither couple could have produced either baby.

3. Which of the following women risk bearing a child with erythroblostis fetalis? a. Mary, who is Rh^+ and whose husband is Rh^-; b. Jill, whose husband is Rh^- and who is Rh^- herself; c. Sally, who is having her first baby. She is Rh^- and her husband is Rh^+ d. Joyce, who is Rh^- but who was once given a transfusion of the Rh^+ blood; her husband is also Rh^+; e. none of the above.

4. A child is born with type AB blood. If its mother is type A, the father is: a. type A, b. type B, c. type AB, d. either (b) or (c) is correct, e. none is correct.

5. What might be the blood type of a husband whose child was type O and whose wife was type B? a. A, b. B, c. AB, d. O, e. a,b,or d.

STUDY QUESTIONS

1. Could you hope to solve this problem without the necessary background knowledge? Will your instructor always supply it in the problem? (p. 413)

2. How come one of the possible squares of the model has only <u>two</u> cells?

3. Which is the better way to do p. 413 do you suppose? The formal Punnet square or this kind of logic?

4. Here is a proposition you could memorize: an AB parent cannot have a type _____ baby!

5. In No. 2 is it necessary to take the ABO information into account at all? (p. 415)

6. Are real life problems usually so nicely set up as in No. 2?

7. What facts would you have to remember from the earlier parts of the course to solve No. 3? (p. 415)

8. Why are there so many possibilities in Nos. 4 and 5? What legal problems might this raise? (p. 415)

III. SEX LINKAGE

LO 24: Given parental genotypes, sexes, and dominances of the genes involved, solve simple sex-linkage problems.
a. when single pairs of alleles are taken into consideration.

PRINCIPLE: In a cross involving sex-linkage, one determines the behavior of the genes by predicting the behavior of the x-chromosome to which they are attached. One must determine (1) the offspring that will receive an x and a y chromosome (2) the offspring that will receive two x chromosomes (3) the origin, whether maternal or paternal of any offspring's x-chromosome(s).

Required information: (1) dominance and recessiveness of the genes involved (2) genotypes <u>together</u> <u>with</u> <u>sex</u> of the parents (3) (in some cases) sex of the offspring.

EXAMPLE 1: Jeannette is heterozygous for color blindness, a recessive, sex-linked trait. She marries Fred, whose vision is normal. Will they have any color blind children? Give the sex.

Given: Color blindness is sex-linked and recessive.
Jeanette is heterozygous for this trait.

Find: Genotypes <u>and</u> <u>sex</u> of children.

Let X^c stand for the abnormal x-chromosome; X^+ for the normal allele. Y is, of course, the Y chromosome.

Jeanette, being female, has two x chromosomes. Since she is heterozygous, her genotype would therefore be $X^c X^+$. Fred being male is $X^+ Y$. He must be X^+, for if he were not, he would be color blind--since a hemizygous gene is always expressed.

What kinds of gametes can each produce?
$$X^C X^+ \longrightarrow X^C \text{ and } X^+$$
$$X^+ Y \longrightarrow X^+ \text{ and } Y$$

Possible combinations are worked out as in a monohybrid cross:

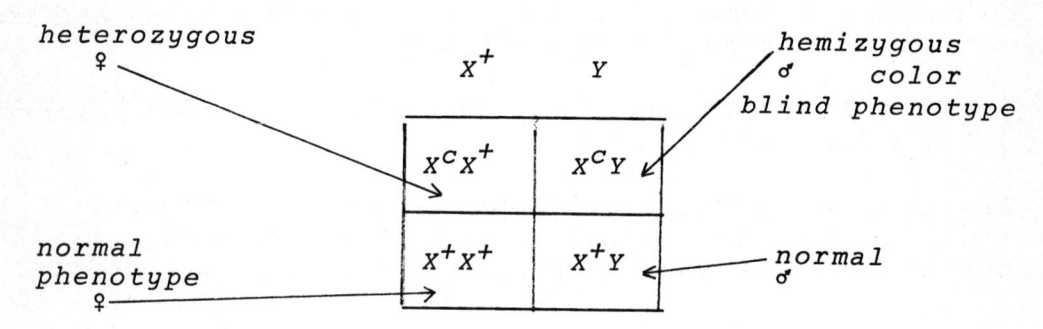

	X^+	Y
	$X^C X^+$	$X^C Y$
	$X^+ X^+$	$X^+ Y$

heterozygous ♀ → $X^C X^+$

hemizygous ♂ color blind phenotype → $X^C Y$

normal phenotype ♀ → $X^+ X^+$

normal ♂ ← $X^+ Y$

The only color blind children, therefore, will be male.

MODEL

Information given _____

_____

Genotypes of parents: $X^- X^-$ and $X^- Y$
(By X^- we mean to indicate a blank for you to fill in)

Possible gametes: (male) _____ and _____
                    (female) _____ and _____

Possible combinations of gametes:

	X^-	Y
X^-	$X^- X^-$	$X^- Y$
X^-	$X^- X^-$	$X^- Y$

Phenotypes (including sex) _____ , _____ ,
_____ , _____ , _____

Conclusion: _____

EXAMPLES FOR YOU TO DO (For help, see p. 425)

1. Betsy is color blind. She marries a normal man. What are the phenotypes and associated sexes of the offspring?

2. Dorie comes from a family in which the boys were hemophilic, and she herself is heterozygous for the trait. She marries a hemophilic man. What are the sexes and associated phenotypes of the offspring?

> LO 24b.: When it is necessary to infer parental sex or genotype.

EXAMPLE 2: Gloria's two sons suffer from Duchenne-type muscular dystrophy, a sex-linked, recessive trait. She and her husband are normal. What is her genotype?

Given: trait is sex linked, recessive boys exhibit it, parents phenotypically +

The boys must be hemizygous for the trait. Each boy received his x-chromosome from his mother. Since their father was normal, his x-chromosome must have been normal.

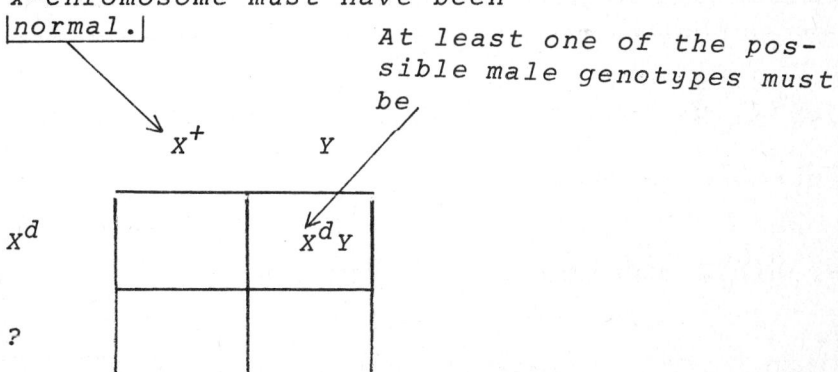

At least one of the possible male genotypes must be

	X^+	Y
X^d		$X^d Y$
?		

Since she is phenotypically normal, she must have had a normal X chromosome. That allows us to complete her side of the Punnett square and if we wish, to work out the genotypes of all the siblings.

	X^+	Y
X^d	$X^d X^+$	$X^d Y$
X^+		

MODEL:

Information given: _____
_____
_____

Genotype of male parent: _____
(easy to infer since any x-linked gene he has will be
expressed.)   gametes: _____

	X^-	Y
X^-		X^-Y
X^-		

Genotype of female parents: _____
(infer from phenotype and the dominance of the genes)

Conclusions: _____
_____

EXAMPLES FOR YOU TO DO:

3. What will be the sexes and associated genotypes of the other siblings in example 2?

4. Judy's mother has normal vision; but Judy is color blind as was her grandfather on her mother's side.
a. What is her mother's genotype?
b. What is her father's phenotype?

> LO 24c: When more than one pair of alleles are taken into consideration.

EXAMPLE 3: Color blindness and hemophilia are both sex-ked recessive traits. A color blind woman whose ther was hemophilic marries a normal man.

a) What is the chance that she is heterozygous for hemophilia?
b) If she is indeed heterozygous for hemophilia what will be the sexes and associated phenotypes of their children?

Given: hemophilia, color blindness sex-linked and recessive

♂ normal
♀ color blind. Brother hemophillic.

Find: a) chance that ♀ is heterozygous for hemophilia.
b) sex, phenotypes of offspring assuming that she is.

First we must try to work out the female genotype. If she had a hemophilic brother, he was X^hY. Therefore:

	X^+	Y
X^h	X^hX	X^hY ← Brother
X^+	XX	

It appears that she has a 50% chance of being heterozygous for the trait. Unless her father is hemophilic, (virtually impossible) it is not certain that she is.

Now we shall proceed on the assumption that she is indeed X^hY. But more than that we must also take her color blindness into account. Since this is a recessive trait, she is X^cX^c. Putting this together, she is $X^{hc}X^c$. (We omit the + sign for practical reasons). Her gametes then are X^{hc} and X^c. His are X and Y.

As you can see, all females will be normal and all males color blind. Half the males will also be hemophilic.

	X	Y
X^{hc}	$X^{hc}X$	$X^{hc}Y$
X^c	X^cX	X^cY

MODEL:

Information given: _____
_____
_____

Genotype of parent? ♂ _____

Genotype of parent? ♀ _____

Conclusion: _____

EXAMPLES FOR YOU TO DO

5. One form of manic depressive mental illness is transmitted as a recessive sex-linked gene. A color-blind manic marries a woman who is heterozygous for hemophilia. What happens?

6. What if she were heterozygous for color blindness and manic. (having had a color blind father) and he normal?

STUDY QUESTIONS

1. What is meant by hemizygous? What traits will be expressed in the hemizygous state?
2. Could Fred heterozygous for this trait?
3. If color blindness is recessive, what, by definition must be the genotype of a color blind female?
4. What is the source of a man's x-chromosome? Is it maternal, paternal or could it be from either parent?
5. Until recently, no one could have a hemophilic father. Why not? How would this affect the sex distribution of hemophilia?
6. Will two sex-linked genes always be found on the same chromosome?
7. Is there any way a girl with Judy's genotype (No. 4) could have been born to a homozygous normal mother?(p.420)

8. Why are both the genes for color blindness and mania on the same x-chromosomes in example No. 4? (p.420)
9. How do you know the abnormal genes are on separate chromosomes? (note that her father is not said to have been a manic).

HINTS AND ANSWERS FOR PART I.

1) RW X RW produces gametes R and W from both parents.

	R	W
R	RR	RW
W	RW	WW

25% WW (White)
50% RW (Pink)
25% RR (Red)

2) SS x Ss produces gametes S and s from both. obviously, 75% of their offspring will be phenotypically normal or nearly so.

3) Ww X Ww is the cross. 25% of the offspring will be ww.

4) Ff X Ff is the cross. 50% of the offspring will be Ff, and an additional 25% FF. The total proportion exhibiting this dominant gene will therefore be 75%

5) RW X WW gives 50% RW (pink) and 50% WW (White). Similarly, RW X RR gives 50% pink, 50% red.

6) ll x Ll. Chance is 50%.

7) Since their second child had the trait, he was homozygous for it (remember it is recessive) and was, if you like, mm. Now he had to get one of his m genes from Chris, and the other from Russ. Therefore, each parent had at least one m gene. But since they are apparently normal people, each must have been Mm. Thus the cross between them was Mm X Mm which you can see, would produce some mm offspring in all likelihood. We have therefore inferred Chris' genotype. Apparently her second husband is genetically normal, MM. He produces gametes with the M gene

only, and she produces them both with the M gene and with the m gene.

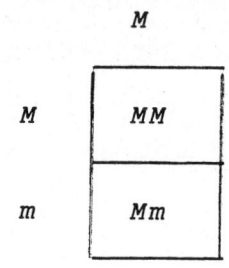

```
          M
      ┌──────┐
  M   │  MM  │
      ├──────┤
  m   │  Mm  │
      └──────┘
```

Obviously, all offspring are phenotypically normal.

HINTS AND ANSWERS FOR PART II

1) Fooled you with this one! Frankie would have to contribute an O gene, which she, being AB does not have. Must be wrong baby.

2) You <u>could</u> make the following crosses (bearing in mind that the Rh and ABO inheritance do <u>not</u> influence one another.)

```
       Wong:  AO X AB
              AA X AB
              RR X RR
              RR X Rr
              Rr X Rr
       White: AB X AB
              rr X rr
```

But what <u>we</u> would do, after having set down the possibilities, is observe that the White family needs less work! Then, we'd ask outselves: "is either of these children a possible White?" The answer is yes -- only the boy is, because he is Rh - negative.
The girl is not, for she would have to have at least one Rh-positive parent. The problem implies that these are the only two possible babies, so assuming that everything is otherwise on the up-and-up, the boy must be young Mr. White, and the answer is (a).

3) For erythroblastosis fetalis to develop, the mother must be Rh-negative and the fetus Rh-positive. Moreover, the condition almost never develops in the first pregnancy. So you should look for a case where the child is <u>possibly</u> going to be Rh^+ and the mother Rh^-. But she must have built up anti-Rh antibodies. Of course the answer is (d).

4) d.

5) e.

HINTS AND ANSWERS FOR PART III.

1) Betsy is $X^c X^c$ (it is recessive)
Thus:

	X^+	Y
X^c	$X^c X^+$	$X^c Y$

2) $X^h X \times X^h Y$

	X^h	Y
X^h	$X^h X^h$	$X^h Y$
X	$X X^h$	XY

3)

	X^+	Y
X^d	$X^d X^+$	$X^d Y$
X^+	$X^+ X^+$	$X^+ Y$

4) Judy is $X^C X^C$. She therefore had to receive X^C from her father, and X^C from her mother. Since her mother is normal she will be X^+X^+ or $X^C X^+$. Since her mother's father was $X^C Y$, mother must be $X^C C^+$. In order to donate an X^C, Judy's dad must be $X^C Y$, phenotypically color blind.

5) $X^{Cm} Y \times X^h X$

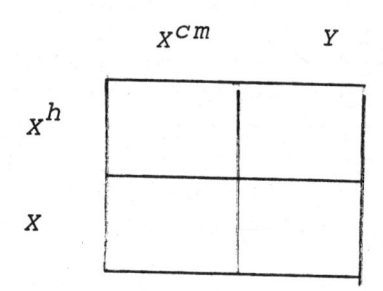

	X^{Cm}	Y
X^h		
X		

6) $X^C X^m \times XY$

PROBLEMS (For help, see p. 428)

1. Mrs. Wiener, newly married, has been tested and shown to be heterozygous for Tay-Sachs disease. Mr. Wiener, however, is free from it. What is the likelihood that a given child will contract this desease?

2. One of Jim's parents is hemophilic, as is he. Which parent?

3. Sally's father is color blind. She marries an apparently normal man. Their child is a color blind albino. What are the genotypes of Sally and her husband with respect to those traits?

4. In the Rh system of blood types, the negative alleles are recessive to the positive. Reba is Rh positive, but her father was negative. She marries Ashford who is Rh negative.
   a. What will be the phenotypes of their children?
   b. Are any of the children likely to suffer from Rh hemolytic disease?

5. Two guinea pigs mate. One guinea pig is white, the other is black. They have five piglets, only one of which is white. The white parents is: a. homozygous, b. heterozygous, c. hemizygous, d. homologous, e. monohybrid.

6. The black parent's genotype is: a. homozygous dominant, because if the dominant gene were heterozygous, it would appear gray. b. heterozygous, because the white baby had to receive a white gene from both parents. c. homozygous, the offspring occur in the wrong proportions for it to be heterozygous. d. there is no way to tell, because the white baby might take after its mother completely, for example, by receiving two white genes from her.

7. If two of the black piglets mate, what will the proportion of phenotypes be among their offspring? a. 25% white 50% gray 25% black, b. 25% black, 75% white, c. 50% black, 50% white, d. 75% black, 25% white e. 100% black.

8. Two red setter dogs were crossed. Out of 40 offsprings, 10 were white. Can you be certain of the genotype of any of these? Which ones? a. parents, b. red offspring, c. white offspring, d. parents and white offspring, e. none of them.

9. Mary is color blind (recessive) and Jim her husband has brown tooth enamel (dominant, sex-linked for purposes of this problem). Which of the children will exhibit both abnormalities? a. the boys, b. the girls, c. one half of the girls, c. one half of the boys, d. one half of both sexes, e. none of them.

10. What will be the phenotypes of Mary's and Jim's grandchildren if one of their daughter's marries a color blind man? a. girls have brown tooth enamel, boys color blind, b. half the girls normal, half color blind; half the boys color blind, half the boys with brown tooth enamel, c. half the children are color blind, half have brown tooth enamel, regardless of sex, d. half the girls are color blind and have brown tooth enamel; other half are color blind only, half the boys have brown tooth enamel; half are color blind.

HINTS AND ANSWERS TO PROBLEMS

1. No chance.   2. In sex-linked inheritance, a son can only receive an X-chromosome from the mother. Since that is the abnormal chromosome, it must have come from his mother. How could he inherit hemophilia from his father?* Any offspring with a paternal X-chromosome will be <u>female</u>.
3. Work out the two separately. He cannot have a color-blind gene, for if he did, being male, he would have the genotype X^cY and would be color blind himself. Obviously she is homozygous for the trait. Assuming Sally to be normally colored, both she and her husband must be heterozygous for the same kind of albinism.
5. (a) Gray should be obviously impossible.
6. (b) If you got (d) as an answer, you are in <u>very</u> serious trouble.
7. The black piglets have to be heterozygous, right? Thus the cross is of the form Aa X Aa $\longrightarrow$ AA, Aa, aa; 1:2:1. Given the dominance of black, that works out to answer (d).
8. (c) What if there were another choice: "all of them"? Could that be right?
9. X^cX^c X X^dY.   Answer is (e).
10. All girl children will be X^bX^c. Remember that brown tooth enamel is a <u>dominant</u> trait. Answer is (d).

---

* Until recently, very few hemophilics grew up to be fathers!